Apostolos Syropoulos and Basil K. Papadopoulos (Eds.)
Vagueness in the Exact Sciences

Also of Interest

Data Science
Time Complexity, Inferential Uncertainty, and Spacekime Analytics
Ivo D. Dinov and Milen Velchev Velev, 2021
ISBN 978-3-11-069780-3, e-ISBN (PDF) 978-3-11-069782-7,
e-ISBN (EPUB) 978-3-11-069797-1

Philosophy of Mathematics
Thomas Bedürftig and Roman Murawski, 2018
ISBN 978-3-11-046830-4, e-ISBN (PDF) 978-3-11-046833-5,
e-ISBN (EPUB) 978-3-11-047077-2

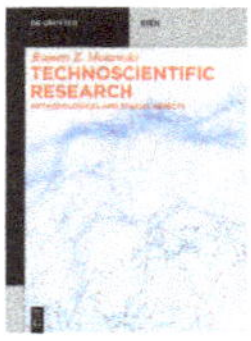

Technoscientific Research
Methodological and Ethical Aspects
Roman Z. Morawski, 2019
ISBN 978-3-11-058390-8, e-ISBN (PDF) 978-3-11-058406-6,
e-ISBN (EPUB) 978-3-11-058412-7

Self-organization of Matter
A dialectical approach to evolution of matter in the microcosm and macrocosmos
Christian Jooss, 2020
ISBN 978-3-11-064419-7, e-ISBN (PDF) 978-3-11-064420-3,
e-ISBN (EPUB) 978-3-11-064431-9

Concepts of Proof in Mathematics, Philosophy, and Computer Science
In: Ontos Mathematical Logic, 6
Edited by: Dieter Probst and Peter Schuster, 2016
ISBN 978-1-5015-1080-9, e-ISBN (PDF) 978-1-5015-0262-0,
e-ISBN (EPUB) 978-1-5015-0264-4

What Reason Promises
Essays on Reason, Nature and History
Edited by: Wendy Doniger, Peter Galison and Susan Neiman, 2016
ISBN 978-3-11-045339-3, e-ISBN (PDF) 978-3-11-045511-3,
e-ISBN (EPUB) 978-3-11-045456-7

Vagueness in the Exact Sciences

Impacts in Mathematics, Physics, Chemistry, Biology, Medicine, Engineering and Computing

Edited by
Apostolos Syropoulos and Basil K. Papadopoulos

DE GRUYTER

Editors
Dr. Apostolos Syropoulos
28th October Str. 366A
671 33 Xanthi
Greece
asyropoulos@gmail.com

Prof. Basil K. Papadopoulos
Democritus University of Thrace
Section of Mathematics
Department of Civil Engineering
Vas. Sofias 12
671 32 Xanthi
Greece
padob@civil.duth.gr

ISBN 978-3-11-070418-1
e-ISBN (PDF) 978-3-11-070430-3
e-ISBN (EPUB) 978-3-11-070437-2

Library of Congress Control Number: 2021938523

Bibliographic information published by the Deutsche Nationalbibliothek
The Deutsche Nationalbibliothek lists this publication in the Deutsche Nationalbibliografie; detailed bibliographic data are available on the Internet at http://dnb.dnb.de.

Cover image: Apostolos Syropoulos
Typesetting: VTeX UAB, Lithuania
Printing and binding: CPI books GmbH, Leck

www.degruyter.com

Preface

Archimedes, (born c. 287 BC, Syracuse, Sicily [Italy]—died 212/211 BC, Syracuse) was a great Ancienet Greek mathematician and inventor. In his treatise entitled *The Sand-Reckoner*[1] [Ψαμμίτης («Ἄμμου Καταμέτρης») in Greek], he calculated the number of sand grains that could be contained in a sphere of the size of our "universe." At that time, the number 10000, which was called miriás (μυριάς in Greek and myriad in Latin), was considered the biggest possible number. Thus Archimedes proposed an extension of the Greek numerical system in order to be able to express huge numbers. This work is really interesting because it shows how we can modify or adapt science in order to incorporate news ideas and findings.

Typically, when we say that something is vague we mean that is not clearly expressed or that it has not a precise meaning. But this means that vagueness is associated with words that do not have a precise meaning or, even worse, we do not know their precise meaning. However, there is a third view according to which the words describe real things and vaguaness is a real thing. In other words, according to this view, the worLd we live is a vague world.

If vagueness (i. e., the possession of borderline cases) is real property of this world, then following Archimedes example, we need to extend our "numerical systems" accordingly. This simply means that we need to examine how *exact sciences* (i. e., sciences for which there are sets of rules to follow or sciences that produce very accurate results) can handle vagueness. There are exact sciences that are intrinsically vague. For example, a distinguished biologist told us that biology is basically a vague science since everything is vague! Also, dermatologists when making a diagnosis of melanoma in the absence of metastasis always make some sort of prediction. Those that favor that vagueness is something that exists only in language give as "yes" or "no" answer. Others that accept vagueness, may conclude that they have found an "atypical Spitz tumor" or "melanocytic tumor of uncertain maligant potential," etc. However, some follow a third way by accepting that they have found a melanoma, a nevus, or something that they do not know what it is.[2] Essentially, the last two approaches are the same as they reject the "yes"/"no" approach, however, we favor the second approach. Naturally, there are other sciences (e. g., mathematics) that are very exact, or at least that is what most people think...

This anthology contains articles contributed by people working with vagueness in various scientific disciplines. The first chapter gives a philosophical view of vagueness. The second chapter describes how we can formulate alternative mathematics by

---

1 Archimedes. (2009). THE SAND-RECKONER. In T. Heath (Ed.), The Works of Archimedes: Edited in Modern Notation with Introductory Chapters (Cambridge Library Collection - Mathematics, pp. 221–232). Cambridge: Cambridge University Press.

2 Kittler H. Binary world/bivalent logic [editorial]. Dermatology Practical and Conceptual 2012;2(2):1. https://doi.org/10.5826/dpc.0202a01

https://doi.org/10.1515/9783110704303-201

using the different expressions of vagueness. In addition, in order to explain certain ideas, the authors use examples from biology, thus explicitly showing that biology is a vague science. Since computing and mathematics are strong tied, the third chapter investigates how we can incorporate vagueness into existing models of computation or introduce new models of vague computing. Statistics is used to "explain" many things, therefore, it should be able to handle vagueness and this is exactly the subject of the fourth chapter. The fifth chapter presents an alternative "interpretation" of quantum mechanics that uses vagueness to explain nature at the subatomic level. In Chapter 6, the authors show why chemistry is basically a vague science. The seventh chapter discusses why medicine is also a vague science. The next chapter takes a look at vagueness in technology. The ninth and last chapter of this anthology examines vagueness from a semiotic point of view. Of course, there are other sciences that are equally vague (e. g., history is another such science) but then we did not want to be exhaustive in our selection of sciences that "look" precise but are not. We just want to make a point: That science should be able to describe vague phenomena, as most phenomena are vague.

At this point, we would like to thank all the authors who have contributed their valuable work to this volume. In addition, we would like to thank Karin Sora, from De Gruyter, who believed in this project and enthusiastically worked with us to make it a book. In addition, we thank all the other people from De Gruyter who helped us. Last, but certainly not least, we thank those people that helped us to finish this project.

Apostolos Syropoulos
Basil K. Papadopoulos
Xanthi, Greece,
April 2021

Contents

List of Contributing Authors

Ken Akiba
Virginia Commonwealth University
Richmond, VA
USA
E-mail: kakiba@vcu.edu

Elena Ciobanu
"Vasile Alecsandri" University of Bacău
Bacău
Romania
E-mail: elena.ciobanu@ub.ro

Pier Luigi Gentili
University of Perugia
Perugia
Italy
E-mail: pierluigi.gentili@unipg.it

Bjørn Morten Hofmann
Norwegian University of Science and Technology
and University of Oslo
Oslo
Norway
E-mail: bjoern.hofmann@ntnu.no

Ioannis Kanellos
IMT Atlantique
Brest
France
E-mail: ioannis.kanellos@imt-atlantique.fr

Fabio Krykhtine
Poli/UFRJ
Rio de Janeiro
Brazil
E-mail: fabio@klam.com.br

Nikos Mylonas
Democritus University of Thrace
Xanthi
Greece
E-mail: nimylona@civil.duth.gr

Basil Papadopoulos
Democritus University of Thrace
Xanthi
Greece
E-mail: papadob@civil.duth.gr

Jarosław Pykacz
University of Gdańsk
Gdansk
Poland
E-mail: pykacz@mat.ug.edu.pl

Apostolos Syropoulos
Xanthi
Greece
E-mail: asyropoulos@yahoo.com

Eleni Tatsiou
Xanthi
Greece
E-mail: etatsiou@gmail.com

Ken Akiba

1 Vagueness from the philosophical point of view

Abstract: This paper offers philosophers' viewpoint on the issues surrounding vagueness in a manner accessible to general readers and scientists. In particular, the paper aims to help answer the following two questions: First, wherein lies vagueness? And second, why do philosophers roundly dismiss fuzzy logic? On the first issue, two kinds of vagueness, worldly vagueness and representational vagueness, are distinguished.

This book is about vagueness in science. However, vagueness has traditionally been even a bigger subject in philosophy than in science. Over the years, but especially in the last 30+ years or so, philosophers have had much heated debate about the nature of vagueness, but the fruits of their discussion have not yet been widely shared in the scientific community. In this paper, I would like to convey, in a manner accessible to scientists and general readers, contemporary philosophers' thoughts on vagueness, which should be helpful to those readers. I will stay away as much as possible from explicitly setting forth my own view; however, the way I present the subject matter will inevitably reflect my own viewpoint.

I will focus on two topics, or two questions. The first question is: Wherein lies vagueness? It seems that vagueness exists somewhere; otherwise, we would not even be talking about it. But where? In particular, does vagueness exist in the world itself, or does it exist only in our representation (perception, language, etc.) of the world? I cannot give a definitive answer to the question, but will try to make clear what the difference amounts to. The second question I will address is: Why do philosophers roundly dismiss fuzzy logic as a way of dealing with vagueness? While the use of fuzzy logic and fuzzy set theory is a widely celebrated treatment of vagueness in natural science and engineering, philosophers are generally very dismissive of their significance. But why? I will answer this question in the second section of this paper.

Throughout the paper, I will use George Klir and Bo Yuan's *Fuzzy Sets and Fuzzy Logic: Theory and Applications* [14] as a point of reference. I chose this book because of its influence and popularity; however, whatever I say about the book is not specific to the book and can easily be generalized.

https://doi.org/10.1515/9783110704303-001

1.1 Wherein lies vagueness?

1.1.1 Introduction

Suppose an astronomer shows you a photo of an alien planet. But this photo is all foggy and blurry, and you cannot detect any solid object in it. What should you think about the photo? There are three possibilities. The first is that it is a crisp picture of the objectively blurry planet. The second is that it is a blurry picture of the objectively crisp planet. The third, the combination of the first two, is that it is a blurry picture of the objectively blurry planet. This distinction between an objectively blurry object and a blurry picture is a crucial distinction philosophers draw that is often neglected by nonphilosophers.

Let's call vagueness of the first kind *worldly vagueness* (or *vagueness in the world*) and vagueness of the second kind *representational vagueness* (or *vagueness in representation relations*). Whether as a result of the worldly vagueness of the planet itself or as a result of the representational vagueness of the photo, there is no doubt that the picture in the photo is vague. For lack of a better name, I will call this vagueness *apparent vagueness* (or *vague appearance*). Then the above three views can be described, respectively, that the apparent vagueness of the picture is as a result of the worldly vagueness of the planet; as a result of the representational vagueness of the photo; and as a result of the combination of those two kinds of vagueness.

In the next two subsections, I will reverse what one may think to be the natural order of presentation, and first give a discussion about what a representation is. I will then give an admittedly sketchy account of the other side of the world-representation divide, objects in the world. You will see why this order of presentation is more convenient. In Subsections 1.1.4 and 1.1.5, I will illustrate, using Klir and Yuan's aforementioned book as an example, how scientists are often confused about the general distinction between what's in the world and what's in a representation of the world, and the distinction between worldly and representational vagueness, in particular. The concept of disquotational truth will be introduced as a possible source of confusion.

1.1.2 Representations

What is a representation? Simply put, a representation is a copy. What is presented in the world (by God?) is *re*presented, that is, presented again, in a representation. Copies, maps, and pictures are representations. Symbols, such as signs and signals, are also representations; for instance, a red light means "Stop!", and smoke means "Fire!" But philosophers are particularly interested in two kinds of representation: *linguistic* and *mental* representations.

Linguistic expressions, sentences in particular, are representations. They try to copy reality. If I said "Lisa is a vegetarian," and if Lisa indeed is a vegetarian, then the sentence succeeded in copying reality; if Lisa is not a vegetarian, the sentence failed to copy reality.

Some mental states, such as perception, beliefs, and thoughts, are representations, too. If I saw that the traffic light in front of me was green, and if it was indeed green, then my perceptual image succeeded in copying reality; if the traffic light was in fact red, the image failed to copy reality.[1] Beliefs and thoughts, unlike perception, are often not accompanied by images; still, if I believed or thought that New York was more populous than London, and if New York is in fact more populous than London, then my belief/thought succeeded in copying reality; if New York isn't, it failed to copy reality. So beliefs and thoughts are also representations.

What are the essential features of a representation? It must have a propositional content, such as that Lisa is a vegetarian, that the traffic light in front of me is green, and that New York is more populous than London, and the content must be true or false, correct or incorrect, or accurate or inaccurate compared to reality. Sentences have *linguistic content*, and the said mental states have *mental content*.

Between linguistic and mental representations, it is usually held that mental representations are primitive and linguistic representations are derivative. Even before we acquired language, we could see, hear, think, and believe things. We invented language to think more clearly and convey our thoughts to others. However, language is much more tangible and easier to deal with than mental states. So philosophers tend to focus on language.

A linguistic expression has both form and content. The study of linguistic content is *semantics* (whereas the study of linguistic form is *syntax* or *grammar*). So the representation relations of linguistic expressions are their *semantic* relations.

A semantic relation connects linguistic expressions, themselves mere sounds or smidges of ink, with the world. Examples are meaning (x means y, where x is a linguistic expression and y is something in the world), expression (x expresses y), denotation (x denotes y), and reference (x refers to y). But the most important semantic relation is *truth* and *falsity*; other relations are important so long as they contribute to truth and falsity of sentences. For instance, if "Lisa" refers to my friend Lisa, "is a vegetarian" stands for the property of *being a vegetarian*, and my friend Lisa indeed has this property, then "Lisa is a vegetarian" will be true; otherwise, it will be false.

---

1 In modern (i.e., 17th–18th century) philosophy, representations were called *ideas*. In particular, John Locke's [16] theory of perception, according to which we perceive the external world not directly but indirectly by obtaining its ideas (i.e., copies), is now called *the representational theory of perception.*

1.1.3 Objects in the world

But I've just stepped over the boundary and started talking about things in the world, such as the individual Lisa and the property of *being a vegetarian*. Let's now think about that side of the world-representation divide, the world, or the objects (things, or entities) in the world. As you will see, I discussed language first because structures of language will help us understand the world.

What sort of things exist in the world? This question is at least as old as Plato and Aristotle, and probably will not be settled anytime soon. However, the contemporary philosophers think that what linguistic expressions represent gives you a good clue about what sort of objects may exist in the world. Think of atomic sentences in language, that is, sentences that do not contain any logical operators or quantifiers, such as "Lisa is a vegetarian," "The traffic light in front of me is green," "New York is more populous than London," and "It is raining." They have two parts: *singular terms* and *predicates*. Singular terms may simply be called "names"; they are names or other designators of individual objects. For example, "Lisa," "the traffic light in front of me," "New York," and "London" are all singular terms. They are all names of (or denote) single, individual, objects, that is, Lisa, the traffic light, New York, and London. (Having parts such as body parts and boroughs does not disqualify them from being individuals, even though those parts themselves may also be considered individuals.)

Singular terms are combined with predicates to make up sentences. Predicates such as "x is a vegetarian," "x is green," and "x is more populous than y" are divided into 1-place predicates, 2-place predicates, 3-place predicates, etc., depending on how many open slots to be filled in with singular terms; so "x is a vegetarian," "x is green," "x talks," etc., are 1-place predicates, "x is more populous than y," "x talks to y," "x loves y," and "x owns y" are 2-place predicates, and "x gives y to z" is a 3-place predicate. Then, by analogy, declarative[2] sentences such as "it is raining" and "it is hot" may be considered 0-place predicates since there is no open slots in them. n-place predicates where $n \geq 2$ are called *many-place* predicates.

In the tradition of Plato, philosophers and linguists consider predicates to stand for *universals*; in particular, 1-place predicates stand for *properties* of individuals, and n-place predicates where $n \geq 2$ stand for *relations* between individuals. 0-place predicates, that is, declarative sentences, stand for *propositions*. So *being a vegetarian*, *being green*, and *talking* are properties, and *being more populous than*, *talking to*, *loving* (or love), ownership, and *giving* are relations, and *that it is raining* is a proposition. When the denotations of predicates are called universals, the denotations of singular terms, individuals, are called *particulars*. Universals are instantiated possibly by many particulars; so, for instance, the property of *being a vegetarian* is instantiated

2 Sentences which are not declarative are interrogative sentences (i. e., questions such as "Is it raining?"), imperative sentences (i. e., orders and requests such as "come here"), and exclamatory sentences (i. e., exclamations such as "Jesus!").

by each individual vegetarian, and the love relation is instantiated by all couples of people in love. Propositions, that is, 0-place relations, are always there without being instantiated by any individual.

When we talk a little loosely, we don't distinguish those different kinds of universals strictly and just say "properties" or "relations" when, in fact, other kinds of universals are included. For instance, at the end of the last subsection I said that truth and falsity are the most important semantic *relations*; however, strictly speaking, they are *properties*, as they are instantiated by single individuals (i. e., sentences). "x is true" and "x is false" are 1-place predicates.

An n-place predicate can be combined with a singular term to create an (n–1)-place predicate; thus, the 2-place predicate "x is more populous than y" can be combined with "London" to create the 1-place predicate "x is more populous than London," and the 1-place predicate "x is a vegetarian" can be combined with "Lisa" to create the sentence (= 0-place predicate) "Lisa is a vegetarian." Analogously, an n-place relation can be satisfied by an individual to create an (n – 1)-place relation. So, the 2-place relation of *being more populous than* can be satisfied by London to create the property (= 1-place relation) *being more populous than London*, and the property of *being a vegetarian* can be satisfied by Lisa to create the proposition (= 0-place relation) *that Lisa is a vegetarian.* There is a perfect analogy between language and reality here.

Let me summarize what we've got so far:

Linguistic expressions	Representation relations	Objects in the world
• Singular terms (names)		• Particulars
• Predicates	denote,	• Universals
– 1-place predicates	stand for,	– Properties
– Many-place predicates	express,	– Relations
• (Declarative) sentences (= 0-place predicates)	is true, etc.	– Propositions

I should note that this is by no means an entirely accurate or complete picture, as I will explain shortly. However, it is good enough at this point.

You may find universals strange objects. Generally, an *abstract object* is an object which does not exist in spacetime; it exists but has no specific spatiotemporal location. Then universals are usually considered abstract objects; they may be instantiated by individuals, which (often) have spatiotemporal locations, but the universals themselves do not have a spatiotemporal location. So, for instance, vegetarianism may be instantiated by each vegetarian, who lives and moves around somewhere in spacetime, but vegetarianism itself has no location. That's Plato's idea.

Many individuals are physical (or material) objects and exist in spacetime. They include minute objects such as fundamental particles, atoms, molecules, and cells, as well as huge objects such as planets, stars, solar systems, and galaxies. Again, some larger individuals may have smaller individuals as their parts; for instance, molecules have atoms as parts, and galaxies have solar systems as parts.

Science also assumes the existence of universals. For instance, being *positively charged* and *being dihydrogen monoxide (i. e., H_2O)* are among numerous scientifically accepted properties.

But there are also more exotic objects assumed in science whose nature is not entirely clear. Mathematical objects, such as sets and numbers, are among them. Science cannot exist without mathematics, but the nature of mathematical objects has never been clarified. Sets and numbers exist independently of us. They are abstract objects, for sure; sets and numbers don't exist in any particular places in spacetime. The set of vegetarians does not exist in any place where each individual vegetarian resides, and number 2 does not exist in any place where there are two individuals. But are they universals, then? It is unclear. The answer partly depends on the exact definition of a universal, but also partly depends on what those mathematical objects are, exactly.

However, one thing is certain, and this will turn out to be important: Sets are different from properties. The set of vegetarians is different from the property of *being a vegetarian*. Properties are more fine-grained than sets. One of the most influential American philosophers, Willard Van Orman Quine (1908–2000) [17], gave the following famous example to make that point. (I don't know if it is a biologically accurate example.) A cordate, by definition, is a creature with a heart. A renate is a creature with a kidney. It turns out, however, that all cordates are renates and all renates are cordates; that is, every creature that has a heart turns out also to have a kidney, and vice versa. So the set of cordates is the same as the set of renates. However, the property of *being a cordate* and the property of *being a renate* are different properties; the former has something to do with hearts and the latter has something to do with kidneys. On the other hand, one and the same property cannot correspond to different sets (at one point of time). For instance, the property of *being a bachelor* and that of *being an unmarried adult male* are one and the same property; so the set of bachelors and the set of unmarried adult males are one and the same set. These examples show that properties are more fine-grained than sets; that is, there are more distinctions among properties than sets.

Indeed, many contemporary philosophers embrace Friedrich Ludwig Gottlob Frege's (1848–1925) [9] idea that linguistic expressions generally stand for two kinds of things, not just the one kind listed on the right side of the table above. For instance, 1-place predicates stand for not just properties but also sets; so, the predicate "x is a vegetarian" *connotes* the property of *being a vegetarian* but also *denotes* the set of vegetarians. "x is a cordate" and "x is a renate" connote different properties but denote one and the same set. Similarly, a sentence connotes a proposition and denotes its truth value, truth or falsity; the singular terms "the original host of the TV show *Apprentice*" and "the 45th President of the United States" connote different concepts but denote the same man, Donald Trump. Frege called connotations *senses* ("Sinn" in German) and denotations *references* ("Bedeutung").

However, this is not a place to get into full-fledged philosophy of language. I said the above to draw a clear distinction between a variety of objects in the world and their representations in language. Nonphilosophers (or even some philosophers) often fail to make this distinction, and confusion and mistakes arise as a result. I will argue in the rest of this section that this is the case in many people's understanding of vagueness.

One feature we can take advantage of in order to distinguish worldly objects from their representations is that the worldly objects exist objectively and independently of us, humans, and our language and mind. So they would exist even without us or language. Thus, even though predicates stand for properties, properties exist without predicates, and even though propositions are meanings of sentences, they exist without being expressed by sentences. However, many people, including many mathematicians and linguists, do not clearly distinguish properties from predicates and propositions from sentences. Confusion ensues as a result, as you will see.

1.1.4 Fuzzy set theory and fuzzy logic

One of the major philosophical issues surrounding vagueness is whether vagueness exists in language or in the world itself. The view that vagueness exists only in language is called the *semantic theory* of vagueness [8, 12, 15], and the view that vagueness exists in the world itself is called the *ontic* or *metaphysical theory* of vagueness [1, 4]. (The latter theory does not need to deny that vagueness exists also in language because language is part of the world – a subtle point we will get back to later.) A large majority of philosophers embrace the semantic theory. For the record, I myself embrace the ontic theory, and don't quite understand why the semantic theory is so popular. But I don't intend to make an argument for the ontic theory in this paper. My point in this paper, rather, is that nonphilosophers don't seem to understand that there is the distinction. Semantic vagueness is a species of representational vagueness I introduced at the beginning of this section. Most semanticists don't quite say this, but I assume that they are using language just as a premier example of representation, and that they don't oppose the idea that vagueness may exist also in other forms of representation, such as perception and other mental states. So the contrast is really between worldly and representational vagueness, as I characterized at the beginning.

As an example of confusion between the two kinds of vagueness, let's consider fuzzy set theory and fuzzy logic, described in detail in Klir and Yuan's book.

The basic idea of fuzzy set theory is that the set-membership relation is not an all-or-nothing relation, as the standard, crisp, set theory describes. According to crisp set theory, either an individual belongs to a set or it doesn't. For instance, either Lisa belongs the set of vegetarians or she doesn't. Fuzzy set theory objects to this idea and introduces *degrees* of membership, which range over real numbers between 0 and 1 inclusive. So, Lisa, who very occasionally eats meat if nothing else is available, may belong to the set of vegetarians to degree 0.96, say.

The basic idea of fuzzy logic is that sentences do not have just one of the two values, truth and falsity, but many values between 0 (falsity) and 1 (truth). According to classical logic, every sentence is either true or false. For instance, "Lisa is a vegetarian" is either true or false. Fuzzy logic objects to this idea and introduces *degrees* in truth values, which range over real numbers between 0 and 1 inclusive. So, "Lisa is a vegetarian" may be true to degree 0.96.

Now, the point I wish to make is this: Klir and Yuan, like many other fuzzy theorists, introduce fuzzy logic as an extension, or a generalization, of fuzzy set theory; but it isn't. Fuzzy set theory and fuzzy logic locate vagueness in completely different places, in the world and in language, respectively. That is, fuzzy set theory is an ontic theory of vagueness whereas fuzzy logic is a semantic theory.

Why? Fuzzy set theory states that the set-membership relation is vague. But, as we just saw, sets exist in the world, independently of us or our representations. So, fuzzy set theory ascribes vagueness to things in the world.[3] In comparison, fuzzy logic states that truth has different degrees. But, as we saw, truth is a semantic relation determined between sentences and the world. So, fuzzy logic ascribes vagueness to semantic, or representational, relations.

One thing which may obscure this radical difference is the fact that Klir and Yuan call what I call sentences "propositions" as in "propositional logic" (although there is no doubt they are referring to sentences by "propositions"). This terminology is innocent in many contexts, but perhaps not here. As I said above, sentences are linguistic expressions whereas propositions are meanings of sentences which, however, exist independently of sentences. But no matter how you call sentences, if you maintain that the truth values of sentences ought to have degrees, then you are ascribing vagueness to representation relations and not the world itself.

1.1.5 Disquotational truth

I think, however, that there is another, more serious and interesting, source of confusion here. I have been saying that truth is a semantic relation (or, more accurately, a semantic property), and semantics is about the representation relations between language and the world. However, there is a concept of truth that is not genuinely semantic; it's called *disquotational truth*. I will call the truth concept that is genuinely semantic and nondisquotational *nondisquotational truth*. When fuzzy logicians say truth has degrees, it is possible that there is confusion between disquotational and nondisquotational truth. It's a complicated topic, but let me explain.

---

3 Incidentally, some semanticists in older generations [6, 15, 20] thought the idea of worldly vagueness incoherent and unintelligible. I suspect some scientists may feel the same way. Fuzzy set theory, however, demonstrates that this view is unfounded, for it definitely ascribes vagueness to some worldly objects, that is, sets.

To begin, let's ask ourselves why we need the truth predicate "*x* is true" in the first place. Well, perhaps we try to describe a certain feature of sentences we ascribe the predicate to, just as we try to describe a certain feature of things by ascribing the predicate "*x* is green." Perhaps we try to say that those sentences somehow correspond to reality. It has been point out, however, that the truth predicate is useful also for purely logical, nondescriptive, purposes.

The truth predicate "*x* is true" satisfies the following schema:

(Disquotation schema) For any sentence *p*,

"*p*" is true if and only if *p*.

Here, *p* is a schematic letter, to be replaced with any declarative English sentence (which does not contain indexical expressions such as "I," "you," "this," "that," "now," "here," "tomorrow," etc., whose denotations change in different contexts).[4] So instances of schema are: "Snow is white" is true if and only if snow is white, "Earth is round" is true if and only if Earth is round, etc. The disquotation schema is so-called because the truth predicate "*x* is true" works like the reverse operation of quotation: for any *p*, if you put quotation marks around it and then attach "is true" at the end, the result is equivalent to *p* itself.

The truth predicate which satisfies the disquotation schema is useful because it allows us to say something about the world without actually describing it. Thus, instead of saying "Snow is white," we can say "'Snow is white' is true." The latter says something about the sentence "Snow is white." Usually, if we put a linguistic expression in quotes, we are talking not about the world but about the expression; by saying "'Plato' is a five-letter word," we are talking not about the philosopher Plato but the expression "Plato." Generally, we won't know anything about the philosopher Plato or snow if somebody says "'Plato' is..." or "'Snow is white' is...." However, if somebody attaches the truth predicate "*x* is true" to "Snow is white," we will know something about the world, that is, the color of snow, thanks to the disquotation schema; we now can say something about the world indirectly by saying something about the relevant sentence. The aforementioned philosopher Quine [18] called this, indirect, way of talking about the world a speech in *formal mode*, and contrasted it with the ordinary way of talking about the world, a speech in the *material mode*, and called the switch from the material mode to the formal mode a *semantic ascent*.

But how is a speech in formal mode useful or necessary? It is useful because we don't need to repeat what (other) people say. Suppose that I believe that Earth is round, and that Lisa says "Earth is round." Then instead of saying "Earth is round" myself,

4 If the relevant sentence contains an indexical expression, the schema does not always hold; for instance, your utterance yesterday of "I am fine today" is true if and only if you were fine yesterday, not if I am fine today.

I can say "What Lisa said is true." Since what Lisa said = "Earth is round," by way of the disquotation schema, I can express my belief about the world, that Earth is round, indirectly. Similarly, by saying "Everything Donald Trump says is true," I can express my belief about the world, that the Mueller investigation was a hoax, that 90 % of corona virus cases are harmless, etc., including things I can't really spell out (since I don't know every assertion Donald Trump makes). In fact, the disquotational truth predicate is not only useful but indispensable in expressions like "Everything ZFC set theory implies is true," as ZFC set theory implies infinitely many things, and thus, it is theoretically impossible to spell out all of them.

This disquotational use of the truth predicate is so rampant that we often don't even realize it. For instance, we often use the word "truth(s)" almost synonymously to "fact(s)." For instance, we sometimes ask, "What's the truth here?" Since what is true is a sentence,[5] we are, on surface, asking for a truth sentence. That sentence might be "Earth is flat." But once we obtain this sentence as a true sentence, then, by the disquotation schema, we will obtain a fact about the world, that Earth is flat. Assertions like "Truth is out there" can be explained in a similar manner.

There are several theories of truth. Perhaps the most popular and historically most influential theory is *the correspondence theory of truth*, according to which truth is a matter of correspondence to reality; a sentence is true if and only if it corresponds to reality. My previous account of the representation relations, although I did not make it explicit back then, is based on this theory. The correspondence theory takes truth to be genuinely semantic and representational. In recent years, however, a rival theory has emerged, *the disquotational (or deflationary) theory of truth* [5, 7, 11]. According to the disquotational theory, the truth predicate and other related semantic predicates (such as "denotes" and "means") is just a disquotational device and does not have any genuine semantic or representational function. The disquotational theory is a radical and *prima facie* implausible theory, denying the representational power of language and mind. Put simply, according to the disquotational theory we don't make copies of reality in language and, in particular, in mind. This is a *prima facie* implausible theory, whatever its final verdict may be, because it does not seem to explain how we have succeeded in surviving in the world. For instance, if we don't make a reasonably accurate visual copy of the world by seeing the world, what would explain our apparent advantage over the blind people? – one must wonder.

1.1.6 Conclusion

However, we don't need to be disquotationalists to appreciate the disquotational function of the truth predicate. We can talk indirectly about the world by declaring sen-

5 In fact, some philosophers argue that in addition to sentences, propositions can be true or false. To avoid an unnecessary complication, I will set aside this possibility.

tences true. Then, perhaps, we may be able to talk indirectly about the vagueness of the world by declaring sentences true only to a certain degree. I wonder if that's what fuzzy logicians are trying to do when they assign degrees to the truth predicate.

Consider the following schema:

(Fuzzy disquotation schema) For any sentence *p*,

"*p*" is true to degree *n* if and only if *p* to degree *n*.

Instances of this schema are: "Lisa is a vegetarian" is true to degree 0.96 if and only if Lisa is a vegetarian to degree 0.96; "Mike is tall" is true to degree 0.25 if and only if Mike is tall to degree 0.25; and "John is bald" is true to degree 0.54 if and only if John is bald to degree 0.54.

If we assume this schema to hold, then we don't have to assume that fuzzy logicians are assigning vagueness to representation relations when they introduce degrees of truth; instead, we can take them as introducing degrees to properties such as *being a vegetarian*, *being tall*, and *being bald*, turning them into *being a vegetarian to degree n*, *being tall to degree n*, and *being bald to degree n*. As I've said earlier, properties are like sets in that they exist in the world itself, although they are more fine-grained than, and thus different from, sets. Then we can take the apparent vagueness of the truth predicate expressed on the left side of the fuzzy disquotation schema merely as a result of the vagueness of worldly properties expressed on the right side. The picture of the planet is fuzzy because the planet is fuzzy.

Degree properties are everywhere in science. Length (*x is n centimeter long*), velocity (*x moves at the speed of n meter per second*), and temperature (*x is n degree Celsius*), to mention a few, are all degree properties. So it would be in line with this common practice to introduce degrees to vague properties.

I wonder if that's what fuzzy logicians are trying to but fail to do because they don't have sufficient awareness and conceptual resources. At the very least, after introducing degrees to set-membership, it is a natural idea to introduce degrees to properties. But fuzzy logicians such as Klir and Yuan never do that; instead, they introduce degrees to truth. They never mention what I call the fuzzy disquotation schema or anything of the sort. So my conjecture remains just that – a conjecture. On the face of it, fuzzy set theory is an ontic theory whereas fuzzy logic is a semantic theory of vagueness. They are based on very different ideas about vagueness.

As I said, my favorite is the ontic theory; however, I don't necessarily object to the introduction of degrees to truth. Once you introduce degrees to vague properties such as *being tall* and *being bald*, and if you think that the truth property is also a vague property, then you can introduce degrees to the truth property, too. This would make you a nondisquotationalist, for the disquotational truth, defined by the original disquotation schema, is precise. But you may embrace the correspondence theory of truth and the genuine representational power of the truth property, but may allow for vagueness in the representation relations including truth. This would be the third position I presented in the planet picture example at the beginning of this section. One somewhat misleading thing about the table I presented earlier is that even though I tried to make clear the distinction between the world, on the one hand, and language and representation relations, on the other, language is indeed part of the world, and representation relations, as relations, are part of the world, too; so vagueness in representation relations are worldly vagueness in a broad sense. The ontic theory of vagueness may accept vagueness in representation relations along with other worldly relations.

But alternatively, the onticist can deny the existence of vagueness in representation relations. This would be the first position in the planet picture example. The onticist can also be a disquotationalist, denying the representational power of the truth predicate and other semantic predicates but embracing the disquotational and fuzzy disquotational schemas. In these cases, there will be apparent vagueness in the truth predicate, but that's only apparent, and all vagueness can be traced ultimately to worldly and nonrepresentational vagueness. So there are several theoretical options for the onticist.

In this section, I did not argue for either the ontic theory or the semantic theory. The most important thing for nonphilosophers to understand, however, is that there is the difference. Once you recognize the difference, then you can make up your mind about which position you support. Without that recognition, everything cannot but be muddled.

1.2 Why do philosophers dismiss fuzzy theory?

1.2.1 Introduction

Since in Section 1.1 I used the term "fuzzy logic" in the narrow sense, contrasting it with fuzzy set theory, I will use the term "fuzzy theory" in this section to denote the whole group of theories including fuzzy set theory and fuzzy logic in the narrow sense, even though the whole group is often called "fuzzy logic" in a broader sense of the term. Even though my point in Section 1.1 was that fuzzy set theory and fuzzy logic are based on different philosophical theories and thus should be clearly distinguished, since that point does not affect my main contentions in this section, I will be a little loose in mixing the two in my narrative.

One notable fact about the investigation into vagueness is that the philosophical approach to vagueness is very different from the scientific approaches. In particular, while fuzzy theory plays a significant – if not the central – role in scientific investigations, it is roundly dismissed and almost completely ignored in philosophy. Why? There are several reasons for this situation, but in what follows I will focus on two main reasons why philosophers dismiss fuzzy theory.

1.2.2 Nihilism about vagueness and the sorites paradox

No doubt, vagueness is a puzzling phenomenon. In Section 1.1, I actually took a very vagueness-friendly approach, assuming that vagueness exists, and asked where it does. There are, however, many researchers, both in philosophy and in science, who don't believe that vagueness really exists. They only believe in precise and crisp objects and properties in the world. They are *nihilists* about vagueness.

The nihilists contend that where there seems to be vagueness, there is in fact only small-scale crispness. Suppose there is a color patch in front of you. The left end of the patch definitely looks red, and the right end definitely looks orange, but the color changes gradually from red to orange in between. The nihilists, however, deny that there are vague properties such as *being red* and *being orange*. According to the nihilists, the surface of the patch consists of very narrow strips each of which emits light waves of a precise wavelength (or a set of precise wavelengths). There is no property of *being red*, and no property of *being orange*, in addition to the precise properties of *emitting light waves of wavelength n nanometers*, for various specific *n*s. Similarly, there are no short people or tall people; there are only persons of specific precise heights. No bald people or hirsute people, only persons with specific numbers of hairs.

Indeed, the famous *sorites paradox* (or the paradox of the heap, as *soros* means "heap" in Greek) can be taken as a *reductio* argument against the existence of vague properties. Using youth as an example instead of heapness, the argument goes as follows: A baby born today is definitely young. A two days old baby is also young, for one day difference in living should make no difference about the baby's youth. A baby three days old is young, for the same reason. ... But if we continue this reasoning, then we end up concluding that an 80-year-old man must also be young. So where did we go wrong? Since it seems undeniable that no single day of life would turn a young man suddenly into a nonyoung man, and that an 80-year-old man is definitely not young (perhaps despite his insistence to the contrary), the only reasonable conclusion seems to be that even a newborn baby is not young, that there is no vague property of *being young*. By the similar consideration, generally there are no vague properties. (So the 80-year-old man is not old, either.)

The main problem with nihilism, however, is that it must deny the existence of many, many ordinary objects and properties, for apparently, vagueness is everywhere. Take the Sun, for example. Its boundary is vague; there is no way to decide whether a particular photon coming out of the Sun at a particular point of time is still a part of the Sun or not. Then, according to nihilism, either the *part of* relation or the object Sun doesn't exist. But if the *part of* relation doesn't exist, then no composite objects (including molecules and atoms) can exist. So it would be actually less damaging to our conceptual scheme if the Sun doesn't exist. So the Sun doesn't exist because its boundary is vague. However, most ordinary objects, animals, plants, people, cells, chairs, tables, etc., have vague boundaries at the microphysical level; they are all like the Sun to some degree. Therefore, those things cannot exist, either. In fact, not only can each of us, humans, not exist because when each of us started existing is vague, the property of *being a human* itself cannot exist because it's a vague property and there are cases in which whether something is a human or not is indeterminable. Three of the papers the (then-)nihilist Peter K. Unger published in 1979 are titled "There are no ordinary things" [23], "Why there are no people" [24], and "I do not exist" [25]. Even a large part of sciences, perhaps except quantum physics, must be a fiction because it involves various vague objects and properties. (But then the quantum world also contains quantum indeterminacy!)

So, if you want to avoid nihilism, you must solve the sorites paradox in some other way. This is where fuzzy theory could, potentially, help. However, philosophers immediately realized that it won't.

1.2.3 Where does degree 1 (or 0) end?

The problem with fuzzy theory is this: Fuzzy set theory gives degrees in the set-membership relation. However, it still draws sharp borderlines between degree 1 (completely in the set) and less than 1, and between degree 0 (completely out of the set) and more than 0. But those borderlines are contradictory to our idea that there are no sharp lines anywhere in vague properties. To be specific, let's take a look at Figure 1, which is a fairly accurate reproduction of a graph on page 20 of Klir and Yuan's book:

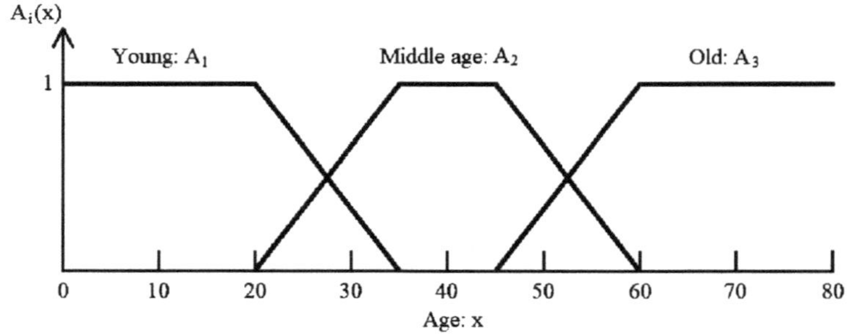

Figure 1: Young, middle-aged, and old.

According to this graph, at age 20 (perhaps on one's 20th birthday), one starts becoming not entirely young, and at age 35, one ends being even slightly young. But how did they decide on those sharp borderlines? Our problem with vagueness was that even though some people are definitely young and some people are definitely not, we cannot find *any* significant borderline in the transition from being young to being nonyoung. Fuzzy theory's alleged solution is similar to the alleged solution by three-valued logic, which gives sentences three values, true, false, and in-between, instead of the usual two values, true and false, that classical bivalent logic gives. It is immediately obvious that three-valued solution is no solution; instead of drawing a sharp borderline between truth and falsity, it only draws two sharp borderlines between truth and in-between, and in-between and falsity. We cannot find those sharp borderlines in reality. But only thing fuzzy logic does is to give a gradation in the in-between area. That's not a solution to the problem for three-valued logic.

Thus, fuzzy logic does not even begin to solve the sorites paradox. It must be said, though, that the sorites paradox has proved a very difficult problem to solve by any philosophical theory, and even a few theories have emerged that concede that there should be sharp borderlines between youth and nonyouth and similar distinctions. For instance, *contexualism* [19, 21] maintains that sharp borderlines exist but that they constantly change depending on the context; that's why we get the impression that there are no sharp borderlines. *The epistemic theory* of vagueness (or *epistemicism* [22, 26]) also argues for sharp borderlines, but tries to explain why we don't, and cannot, know where they are. Even the aforementioned semantic theory, although it usually downplays this point, seems to accept the existence of sharp borderlines in the world by emphasizing that vagueness exists only in our language.

However, the advocates of those theories of vagueness at least recognize the problem and try to explain it away, even though none has emerged as the consensus solution to the sorites paradox. In contrast, fuzzy theorists do not seem to even recognize the problem and make no effort to address it. It is this lack of recognition, not just its inability to solve the paradox, that has repelled philosophers from fuzzy theory.

1.2.4 Penumbral connections and logical truths

That's the first of the two problems with fuzzy theory that I wanted to mention. However, if the first problem were the only major problem with fuzzy theory, philosophers perhaps would have been more kind to it. It is the second problem, to which I will now turn, that has nailed the coffin for fuzzy theory.

The problem concerns the truth degrees of complex sentences, that is, sentences that contain logical connectives. According to fuzzy theory, when some sentences p and q receive some values $[[p]]$ and $[[q]]$ between 0 and 1, the negation $\neg p$ ("not p"), the conjunction $p \wedge q$ ("p and q"), and the disjunction $p \vee q$ ("p or q") will receive the following values:

1. $[[\neg p]] = 1 - [[\phi]]$;
2. $[[p \wedge q]] = \min([[p]], [[q]])$;
3. $[[p \vee q]] = \max([[p]], [[q]])$.

These values, however, are so implausible that it is difficult to take them seriously. Consider, for instance, p "Lisa is young," q "Lisa is middle-aged," and r "Lisa is old."

Suppose Lisa is 27 years and 6 months old. Then, according to Figure 1, $[[p]] = 0.5$ and $[[q]] = 0.5$. Thus the value of $p \vee q$, "Lisa is either young or middle-aged," will be 0.5 (i. e., $[[p \vee q]] = 0.5$), too. That's wrong; it should be 1. True, Lisa is somewhere between young and middle-aged; but she is definitely young *or* middle-aged. Compare this with $p \vee r$, "Lisa is either young or old." Since $[[r]] = 0$, $[[p \vee r]] = 0.5$. This is a plausible degree, as "Lisa is either young or old" is only as likely to be true as "Lisa is young." Also compare the first disjunction with the disjunction $p \vee s$, where s is "Lisa is intelligent." Suppose $[[s]] = 0.5$. Then $[[p \vee s]] = 0.5$. So $[[p]] = [[q]] = [[s]] = [[p \vee q]] = [[p \vee s]] = 0.5$. But that $[[p \vee s]] = 0.5$ seems much more reasonable than that $[[p \vee q]] = 0.5$ because Lisa's being intelligent is independent from Lisa's being young whereas Lisa's being middle-aged isn't.

What these examples show is that not only the above valuations for complex sentences are inadequate, but inadequate in a significant way: We cannot take the value of the disjunction simply as a function of the values of its disjuncts; the contents of the disjuncts matter. Kit Fine [8] called a connection like that between "Lisa is young" and "Lisa is middle-aged" a *penumbral connection* (penumbra = borderline area). Fuzzy logic, then, is inadequate because it ignores penumbral connections.

There is no easy fix of this problem in fuzzy theory. Classical bivalent logic is *truth-functional*; that is, the truth value of a complex sentence is a function of the truth values of its constituents, regardless of their contents. Fuzzy logic is a generalization of classical logic in that it has degrees in [0, 1] instead of just two truth values, 0 and 1. Consequently, it is *degree-functional*: the degree of a complex sentence is a function of the degrees of its constituents, regardless of their contents. The degree-functionality is an essential feature of fuzzy logic. The above example shows that this feature is untenable.

This problem becomes magnified when we think about classical logical truths such as $p \vee \neg p$ "Lisa is either young or not young." This sounds like a tautology; so it should be of degree 1. However, according to the above valuation scheme, it is of degree 0.5, for both disjuncts are of degree 0.5. That means that "Lisa is either young or not young" is as likely to be false as it is true. That just doesn't seem right. Again, the content of $\neg p$, that it is the negation of p, is ignored in the valuation.

Generally, many classical inferences and classical logical truths break down in fuzzy logic even though fuzzy logic is supposed to be a generalization of classical logic. In fact, the implication relation of fuzzy logic is known to be very complicated [10].

Philosophers are generally open-minded about the use of nonstandard logics; indeed, various nonstandard logics have been proposed to deal with vagueness and solve the sorites paradox. However, fuzzy logic's violation of penumbral connections makes philosophers feel that some essential insight is missing in its approach. Combined with its lack of insight into the matter discussed in the last subsection, philosophers generally find fuzzy theory philosophically uninteresting and not worthy of consideration.

In fact, philosophers in general, and I in particular, find fuzzy theory's practical success puzzling, given that it assigns such implausible degrees to complex sentences. I am hoping that other papers in this volume give some answers to this puzzlement.

To conclude: In this paper, I have given philosophers' viewpoint on the issue of vagueness. In Section 1.1, I drew the distinction between worldly vagueness and representational vagueness and illustrated, using Klir and Yuan's book as an example, how scientists may not be clear about the distinction. In Section 1.2, I explained why philosophers are generally dismissive of fuzzy theory as a solution to the philosophical puzzles about vagueness. I spelled out two aspects in which fuzzy theory doesn't do well: First, it doesn't seem to even realize that it is drawing sharp borderlines between the degree 1 area, the in-between area, and the degree 0 area. And second, fuzzy logic is degree-functional and doesn't deal with the problem of penumbral connections.

Needless to say, there are many other issues about vagueness philosophers often discuss and I didn't even touch on. However, I chose my topics with an eye on the expected readership of this volume. Many of the other issues are either outside of general interest or difficult to explain in the limited space assigned. The interested reader should consult the books recommended below.

1.3 Philosophical literature on vagueness

For the reader's convenience, I will recommend some useful philosophy books on vagueness here.

Among monographs, Timothy Williamson [26] and Rosanna Keefe [12] have wide coverage of various theories of vagueness in addition to presenting their own views; however, their coverage of the ontic theory is thin and outdated.

Among anthologies, Keefe and Peter Smith [13] contain important and historically influential papers on vagueness; however, again, its coverage of the ontic theory is thin.

For the ontic theory, Akiba and Ali Abasnezhad [3] is the only book currently available.

Stanford Encyclopedia of Philosophy (https://plato.stanford.edu/) is a generally useful online encyclopedia in philosophy and contains some articles on vagueness, although I have no special recommendation.

All these books and articles are for specialists and philosophy students and rather difficult for nonphilosophers, general readers, and even scientists. If you want to gain some background knowledge about philosophy, especially the things discussed in Section 1.1, Akiba [2] may help.

Bibliography

[1] K. Akiba, Vagueness in the world, Noûs 38 (2004), 407–429.
[2] K. Akiba, The Philosophy Major's Introduction to Philosophy: Concepts and Distinctions, Routledge, 2021.
[3] K. Akiba and A. Abasnezhad, Vague Objects and Vague Identity: New Essays on Ontic Vagueness, Springer, 2014.
[4] E. Barnes and J. R. G. Williams, A theory of metaphysical indeterminacy, in: Oxford Studies in Metaphysics, K. Bennett and D. W. Zimmerman, eds, Vol. 6, Oxford University Press, 2011, pp. 103–148.
[5] R. Brandom, Making It Explicit, Harvard University Press, 1994.
[6] M. Dummett, Wang's paradox, Synthese 30 (1975), 301–324.
[7] H. Field, Deflationist views of meaning and content, Mind 103 (1994), 249–285.
[8] K. Fine, Vagueness, truth and logic, Synthese 30 (1975), 265–300.
[9] G. Frege, Über Sinn und Bedeutung, Z. Philos. Philos. Kritik 100 (1892), 25–50.
[10] P. Hájek, Metamathematics of Fuzzy Logic, Kluwer, 1998.
[11] P. Horwich, Truth, 2nd edn, Oxford University Press, 1999.
[12] R. Keefe, Theories of Vagueness, Cambridge University Press, 2000.
[13] R. Keefe and P. Smith, Vagueness: A Reader, MIT Press, 1999.
[14] G. J. Klir and B. Yuan, Fuzzy Sets and Fuzzy Logic: Theory and Applications, Prentice Hall, 1995.
[15] D. Lewis, Many, but almost one, in: Ontology, Causality, and Mind: Essays in Honour of D. M. Armstrong, K. Campbell, J. Bacon and L. Reinhardt, eds, Cambridge University Press, 1993, pp. 23–38.
[16] J. Locke, An Essay Concerning Human Understanding, 1690.
[17] W. V. O. Quine, Two dogmas of empiricism, Philos. Rev. 60 (1951), 20–43.
[18] W. V. O. Quine, Philosophy of Logic, 2nd edn, Harvard University Press, 1986.
[19] D. Raffman, Vagueness without paradox, Philos. Rev. 103 (1994), 41–74.
[20] R. M. Sainsbury, Why the world cannot be vague, South. J. Philos. 33 (1994), 63–82.
[21] S. Shapiro, Vagueness in Context, Oxford University Press, 2006.
[22] R. A. Sorensen, Blindspots, Oxford University Press, 1988.
[23] P. Unger, There are no ordinary things, Synthese 41 (1979), 117–154.
[24] P. Unger, Why there are no people, Midwest Stud. Philos. 5 (1979), 411–467.
[25] P. Unger, I do not exist, in: Perceptions and Identity, G. Macdonald, ed., Cornell University Press, 1979, pp. 235–251.
[26] T. Williamson, Vagueness. Routledge, 1994.

Apostolos Syropoulos and Eleni Tatsiou

2 Vague mathematics

Abstract: If vagueness is a fundamental property of our cosmos, then we should be able to describe our vague cosmos using an appropriate language. Since we are using mathematics to describe our cosmos, we should be able to define vague mathematics. Basically, there are three different expressions of vagueness, and we demonstrate how these expressions can be used to define alternative mathematics based on vagueness.

2.1 Introduction

Every attempt to understand and/or to describe the world cannot be successful without the use of mathematics and philosophy, or as Gottfried Wilhelm Leibniz has famously said:

> Without mathematics we cannot penetrate deeply into philosophy. Without philosophy we cannot penetrate deeply into mathematics. Without both we cannot penetrate deeply into anything.

Provided that vagueness is a fundamental property of the world we live, which is a basic assumption of this book, we need a philosophy of vagueness and vague mathematics in order to be able to penetrate deeply into anything. An exploration of a philosophy of vagueness has been given in [1] and elsewhere in this book. The purpose of this chapter is to describe alternative mathematics that can be used to model and/or describe vagueness and, consequently, our world. Since mathematics is the epitome of exactness and rigor, it seems that vague mathematics is an oxymoron. On the contrary, there is no oxymoron since mathematics (set theory, category theory, or type theory) is a language that we use to describe the world around us. And if the world is vague, the language should be able to describe it correctly.

There are at least three different expressions of vagueness [13], which could be used to define alternative mathematics.[1] These expressions are:

Many-valued and fuzzy logics Borderline statements are assigned truth-values that are between absolute truth and absolute falsehood.

Supervaluationism The idea that borderline statements lack a truth value.

Contextualism The truth value of a proposition depends on its context (i. e., a person may be tall relative to American men but short relative to NBA players).

1 We wrote “at least” since paraconsistency can also be used to describe vagueness; nevertheless, there are no paraconsistent mathematics although we can find inconsistencies in mathematics (see [10]).

https://doi.org/10.1515/9783110704303-002

The first expression has been used to define *fuzzy mathematics* (e. g., see [14]). Similar to fuzzy sets are *rough sets*. The idea is that a vague entity (e. g., a set) is described by two crisp sets: its *upper* and its *lower* approximations. For the second expression, there is some preliminary work [16] that may eventually lead to the formulation of some alternative mathematics. As for the third expression, there are scholars [8, 9] who had argued that mathematics is context-dependent.

Plan of the chapter

We start by presenting the basic ideas of fuzzy mathematics and by discussing some more advanced ideas. This is followed by a short presentation of rough sets. Then we will discuss how one can start defining vague mathematics based on supervaluationism. Next, we will discuss the idea that mathematics is actually context-dependent.

2.2 Fuzzy mathematics

In 1965, Lotfi Asker Zadeh [21] introduced his *fuzzy sets*,[2] which are an extension of ordinary sets. Unfortunately, Zadeh believed that *fuzziness* and *vagueness* are different things, something that is not correct. In fact, fuzzy sets describe the degree to which entities have a particular property, which is exactly what vagueness is about. Formally, we have the following.

Definition 2.2.1. Let X be a *universe* (i. e., an arbitrary set). A *fuzzy subset* A of X, is characterized by a function $A : X \to [0,1]$, which is called the *membership function*. For every $x \in X$, the value $A(x)$ is called a *degree to which element* x *belongs to the fuzzy subset* A.

In the previous definition, we stated that a fuzzy subset is *characterized* by a function and not it *is* just a function simply because Zadeh used this term. However, we know that $\mathcal{P}(X)$, the powerset of X, is isomorphic to 2^X, the set of all functions from X to $\{0,1\}$, and similarly, $\mathcal{F}(X)$, the set of all fuzzy subsets of X, is isomorphic to the set of all fuzzy characteristic functions, $[0,1]^X$. Therefore, there is no need to distinguish between “is” and “characterized!” For reasons of simplicity, the term “fuzzy set” is preferred over the term “fuzzy subset.”

The basic operations of *crisp* sets (that is how ordinary sets are called in the literature of fuzzy set theory) have their fuzzy counterparts:

Definition 2.2.2. Assume that $A : X \to [0,1]$ and $B : X \to [0,1]$ are two fuzzy sets of X. Then,

2 In 1967 Dieter Klaua [6], while working independently from Zadeh, published his work on *many-valued sets*, which are in a sense equivalent to Zadeh’s fuzzy sets.

- their *union* is

$$(A \cup B)(x) = \max\{A(x), B(x)\};$$

- their *intersection* is

$$(A \cap B)(x) = \min\{A(x), B(x)\};$$

- the *complement*[3] of A is the fuzzy set

$$A^{\complement}(x) = 1 - A(x), \quad \text{for all } x \in X;$$

- their algebraic product is

$$(AB)(x) = A(x) \cdot B(x);$$

- A is a *subset* of B, denoted by $A \subseteq B$, if and only if

$$A(x) \leq B(x), \quad \text{for all } x \in X;$$

- the *scalar* cardinality[4] of A is

$$\text{card}(A) = \sum_{x \in X} A(x)$$

- the fuzzy *powerset* of X (i. e., the set of all ordinary fuzzy subsets of X) is denoted by $\mathscr{F}(X)$.

Example 2.2.1. Suppose that X is the set of all pupils of some class. Then we can construct the fuzzy subset of tall pupils and the fuzzy subset of obese pupils. Let us call these sets T and O, respectively. Then $T \cup O$ is the fuzzy set of pupils that are either tall or obese while $T \cap O$ is the fuzzy subset of pupils that are tall and obese.

Example 2.2.2. Biologists classify living organisms into species. But what are species? The following definition is borrowed from [17].

> A maximally large group of animals, such that healthy young specimens of the right age and sex are able in principle to produce fertile offspring under favorable circumstances that occur naturally.

---

3 Recall that if A is a subset of a universe X, the set of all those elements of X that do not belong to A is called the *complement* of A. This is denoted by $A^{\complement}$.

4 The cardinality of a set A is denoted by card(A) and it is equal to the number of elements of A.

This definition is supposed to handle many and different cases, nevertheless, there are many cases where it fails. In particular, Kees van Deemter [17] discusses why this definition fails in the case of Ensatina salamanders. There are six subspecies and there are subspecies for which it is possible to interbreed two individuals of opposite sex that belong different subspecies. However, the general rule is that two individuals of opposite sex that belong to different subspecies cannot interbreed. Since these salamanders live in a specific area in California's Central Valley, one would expect that interbreeding should be possible. Now consider a more general problem that will make clear what really happens here. Suppose that Julia is able to trace back her ancestry up to 250,000 generations (roughly 7 million years). Theoretically, Julia should be able to inbreed with her father, her grandfather, etc. However, at some time t it would be not possible for her to inbreed mainly because the male ancestor will have many biological differences with his female descendant.[5] Let us now go forward in time and let us start with Julia's first ancestor. If we are careful enough, we may end up with a chimpanzee living somewhere in Gambia. Why? Because the creatures that lived 7 million years ago were our common ancestors. And from all of these we can conclude that chimpanzees are humans! A paradox that van Deemter discusses at length in his book.

Example 2.2.3. Another example [4, 5] that shows that species is a vague concept is about bacteria species. In particular, it has been shown that the "species" *Neisseria meningitidis* and *Neisseria lactamica* are not perfectly distinguished by the sequences of seven housekeeping genes. However, *Neisseria gonorrhoeae* is clearly distinct. In different words, when we spot a *Neisseria* bacterium we can assert to some degree whether it belongs to *Neisseria meningitidis* or *Neisseria lactamica*, but we can definitely say whether it belongs to *Neisseria gonorrhoeae*. Thus if we have a population of bacteria we could form fuzzy sets that would describe to which species they belong.

In crisp mathematics (i. e., ordinary mathematics), we start with sets and use them to build all sorts of mathematical structures. And this is exactly how we do define topological spaces and other structures. Since the aim of this text is to describe how we define alternative mathematics, and consequently, alternative mathematical structures, we will show how one can go from crisp topological spaces to fuzzy topological spaces. Let us start with the standard definition of topological spaces.

Definition 2.2.3. A *topological space* is a set X together with a collection of subsets of X, τ, called *open* sets that have the following properties:

5 Obviously, this looks like the sorites paradox that goes as follows: 10,000 grains of wheat is a heap of wheat. If 10,000 grains of wheat is a heap of wheat, then 9,999 grains of wheat is a heap of wheat. 9,999 grains of wheat is a heap of wheat, then 9,998 grains of wheat is a heap of wheat. Going further down, we can conclude that 1 grain of wheat is a heap of wheat!

1. the intersection of two open sets is open;
2. the union of any collection of open sets is open; and
3. the empty set $\emptyset$ and the whole space X are open.

In addition, a subset $C \subset X$ is called *closed* if its complement $C^{\complement}$ is open. The collection τ is called a *topology on* X.

Fuzzy topological spaces have been introduced by C. L. Chang [3]. Now it is interesting to see how the crisp definition is transformed into a new one.

Definition 2.2.4. A fuzzy topology is a family τ of fuzzy sets in X that has the following properties:
1. $\emptyset, \chi_X \in \tau$, where $\emptyset(x) = 0$ for all $x \in X$ and χ_X is the characteristic function of X (i. e., $\chi_X(x) = 1$, for all $x \in X$);
2. if $A, B \in \tau$, then $A \cap B \in \tau$; and
3. if $A_i \in \tau$ for each $i \in I$, then $\bigcup_{i \in I} A_i \in \tau$.

All members of τ are called *open fuzzy sets* and a fuzzy set is *closed* if and only if its complement is open.

In simple words, a collection of fuzzy subsets of some crisp set X is a fuzzy topology if these fuzzy subsets have the corresponding properties of the open sets of a topological space. This means that it is easy, if not straightforward, to define fuzzy mathematical structures. The truth is that in many cases, these obvious extensions do not yield interesting structures and we need to add some extra constraints or additional properties to make them really interesting. And this is why there are alternative definitions of fuzzy topological spaces. For example, one can say that open sets are open to some degree. Thus if A is open to some degree d, then $A^{\complement}$ is closed not with degree $1 - d$ but with degree d'.

2.3 Rough sets

Rough sets were introduced by Zdzisław Pawlak [11] in 1982. In order to introduce the notion of a rough set, it is necessary to introduce some mathematical prerequisites. Assume that X is a universe (i. e., a set) and R is an equivalence relation[6] on X. An equivalence relation R partitions X into blocks of R-related elements, which are called equivalence classes. Typically, the equivalence class of an element a is denoted by $[a]$

6 An equivalence relation R on a set X is a subset of $X \times X$ (i. e., $R \subseteq X \times X$) such that $a\ R\ a$ for all $a \in X$, $a\ R\ b$ implies $b\ R\ a$ for all $a, b \in X$, and $a\ R\ b$ and $b\ R\ c$ imply $a\ R\ c$ for all $a, b, c \in X$.

and is defined as the set

$$[a] = \{x \mid (x \in X) \wedge (a\,R\,x)\}.$$

The pair (X, R) is called an *approximation space* and R is called an *indiscernibility* relation. When $x, y \in X$ and $x\,R\,y$, x and y are indistinguishable in (X, R). The set of all equivalence classes X, which is denoted by X/R and is called a *quotient* set, forms a partition of X. Any element $E_i \in X/R$ is called an *elementary set*.

If $A \subseteq X$, the *upper approximation*, $\overline{A}$, and the *lower approximation*, $\underline{A}$, of A are defined as follows:

$$\overline{A} = \bigcup_{E_i \cap A \neq \emptyset} E_i = \{x \mid (x \in X) \wedge ([x]_R \cap A \neq \emptyset)\}$$

$$\underline{A} = \bigcup_{E_i \subseteq A} E_i = \{x \mid (x \in X) \wedge ([x]_R \subseteq A)\}$$

In words, the upper approximation of A is the union of R-equivalence classes whose intersection with A is nonempty while the lower approximation of A is the union of R-equivalence classes contained in A. An element $x \in X$ definitely belongs to A when $x \in \underline{A}$. The same element possibly belongs to A when $x \in \overline{A}$. The *boundary* Bnd(A) of A is the set difference $\overline{A} \setminus \underline{A}$. A set $A \subseteq X$ is said to be *definable* in (X, R) if and only if Bnd$(A) = \emptyset$.

Definition 2.3.1. Given an approximation space (X, R), a rough set in it is a pair (L, U), such that L and U are definable sets, $L \subseteq U$, and if any R-equivalence class is a singleton $\{s\}$ and $\{s\} \in U$, then $\{s\} \in L$.

Alternatively, one could say that these two sets define the rough set as they specify its lower and upper boundaries. Figure 1 depicts the boundary of a set A, its lower approximation, and, essentially, its upper approximation. The *accuracy of approximation* [12]

$$\alpha_R(A) = \frac{|\underline{A}|}{|\overline{A}|},$$

characterizes the accuracy of the representation. Clearly, the number $\alpha_R(A)$ is greater or equal to zero and less than or equal to one. In the particular case where $\alpha_R(A) = 1$, set A is crisp with respect to R.

The membership function for rough sets is defined by employing the relation R as follows:

$$\mu_A^R(x) = \frac{|A \cap [x]|}{|[x]|}.$$

It is obvious that $0 \le \mu_A^R(x) \le 1$.

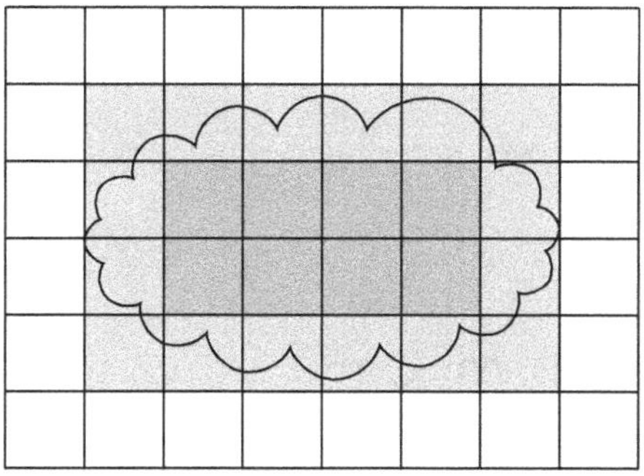

Figure 1: A graphical representation of a rough set. The cloud-like shape shows the boundary of a set A, while "crisp" sets are defined with squares. From the colored squares, those that are painted with the darker color form the lower approximation of A and those that are painted with the lighter color form the difference between the upper and the lower approximations.

In the previous section, we discussed how we can define a fuzzy topological space, and thus we demonstrated that it is possible to create fuzzy mathematics. Naturally, it would be nice to do the same with rough sets. Although this is not that straightforward, still it is possible. For example, Wu et al. [20] have shown how we can construct some sort of rough topological space. Their work is based on the assumption that given a set C and the Euclidean distance between its members, we can define a metric space. Let us first introduce some useful definitions.

Definition 2.3.2. Let $K = (X, R)$ be an approximation space. Assume that $A = (\underline{A}, \overline{A}) \subseteq X$. Let B_1 and B_2 be the subsets that correspond to the metric spaces $\underline{A}$ and $\overline{A}$, respectively. $b_1 \in B_1$ is called an inner point of B_1 inside $\underline{A}$ if b_1 has a spherical neighborhood $\mathcal{B}(b_1, \epsilon_1) \subseteq B_1$; correspondingly, $b_2 \in B_2$ is called an inner point of B_2 inside $\overline{A}$ if b_2 has a spherical neighborhood $\mathcal{B}(b_2, \epsilon_2) \subseteq B_2$. The set that consists of all inner points of B_1 inside $\underline{A}$ is called the interior of B_1 inside $\underline{A}$; similarly, the set that consists of all inner points of B_2 inside $\overline{A}$ is called the interior of B_2 inside $\overline{X}$. These sets are denoted by $\operatorname{Int} B_1$ and $\operatorname{Int} B_2$, respectively. B_1 is called an open set of $\underline{A}$ if $B_1 = \operatorname{Int} B_1$ and B_2 is called an open set of $\overline{A}$ if $B_2 = \operatorname{Int} B_2$. Finally, if B_1 and B_2 satisfy $B_1 \subseteq \underline{A} \subseteq B_2 \subseteq \overline{A}$, $B = (B_1, B_2)$ is called an open set of A.

Using these definitions and some auxiliary results, they have managed to prove the following result.

Theorem 2.3.1. *Assume that $K = (X, R)$ is an approximation space. Suppose that $A = (\underline{A}, \overline{A}) \subseteq X$. Let $\mathcal{T} = (\mathcal{T}_1, \mathcal{T}_2)$ be a family of open sets that consists of all open sets of A. Then $\mathcal{T}$ satisfies the following:*
1. *$A \in \mathcal{T}$ and $\emptyset \in \mathcal{T}$.*
2. *If $O_1, O_2 \in \mathcal{T}$, then $O_1 \cap O_2 \in \mathcal{T}$.*
3. *If $O_i \in \mathcal{T}$, then $\bigcup_i O_i \in \mathcal{T}$, $i \in \{1, 2, \ldots\}$, that is, the union of any number of open sets that are members of $\mathcal{T}$ still belongs to $\mathcal{T}$.*

Here, $\mathcal{T}_1$ and $\mathcal{T}_2$ are families of open sets that consist of all open sets of metric space $\underline{A}$ and $\overline{A}$, respectively.

To put it simply, they have shown how one can construct a rough topology starting from an approximation space.

2.4 Supervaluational mathematics

According to Van Bendegem [15], the use of truth values that belong to $\{0, 1\}$, or to $[0, 1]$, or to any other set is quite problematic. For instance, how can we estimate that an element x belongs to a fuzzy subset A with degree equal to 0.34567? Of course, it is quite possible that such a number is the result of some complicated process, still let us assume that this number make not much sense. On the other hand, we all have been asked to rate certain things and we are prompted to choose a number from a range of integer numbers. Thus we do not really think that using numbers is problematic at all. However, we will stop arguing about this problem and we will turn our attention to Van Bendegem's alternative mathematics.

The essence of Van Bendegem's approach is to use a strategy in order to define a supervaluational interpretation of vague predicates. The first step in his strategy is to define models of formal languages (note that these models are a bit different from the ones in the general literature).

A "model" M of a formal language L is used to interpret it. M is a triple (D, I, v), where D is a nonempty set called the *domain*, I is an *interpretation mapping*, and v is a *valuation* function that maps sentences A onto $\{0, 1\}$ according to some set of semantical rules. The mapping I assigns to each constant symbol c of L and each variable symbol of L to D and to each n-ary predicate symbol P a set of n-tuples of elements of D. In simple words, a constant symbol or a variable becomes D and an n-ary predicate symbols becomes a subset of D^n. Bastiaan Cornelis van Fraassen [18] introduced *supervaluations* and defined them as follows:

Definition 2.4.1. A supervaluation over a model is a function that assigns T (F) exactly to those statements assigned T (F) by all the classical valuations over that model.

A better definition by the same author appeared in [19].

Definition 2.4.2. A valuation s is a supervaluation for a language L if and only if there is a nonempty set K of admissible valuations (i. e., valuations of its syntax) for L such that, for all sentences A of L,

$$\begin{aligned}
s(A) = T &\quad \text{iff } v(A) = T \text{ for all } v \in K,\\
s(A) = F &\quad \text{iff } v(A) = F \text{ for all } v \in K,\\
s(A) &\quad \text{is not defined otherwise.}
\end{aligned}$$

In the second step, Van Bendegem defined the set of all valuations by first defining the various interpretations. Given an n-ary predicate P, then $I^+(P) \subseteq D^n$. The set $I^+(P)$ is called the *positive* extension and the set $D^n \setminus I^+(P) = I^-(P)$ the *negative* extension. This means that for a predicate P the full interpretation $I(P)$ is the pair $(I^+(P), I^-(P))$ such that $I^+(P) \cup I^-(P) = D^n$ and $I^+(P) \cap I^-(P) = \emptyset$.

Next, it is necessary to build a set $\mathbf{M}$ of models $M_1, M_2, \ldots, M_n$ such that, if, in model M_i a predicate P has an interpretation $I_i(P) = (I_i^+(P), I_i^-(P))$, and such if $i < j$, then $I_i^+(P) \subseteq I_j^+(P)$, and hence, $I_j^-(P) \subseteq I_i^-(P)$. Note that the resuly is a "nested" set of models.

The final step is to apply supervaluation to the set $\mathbf{M}$ to obtain truths, falsehoods, and undetermined cases.

Now we need to show how this approach actually works. We will restrict ourselves to a simple language that can be used to express statements about numbers that belong to $\mathbb{N}_0$ (i. e., the set of natural numbers). Its model is the triple (D, I, v), where:

1. D is the set of natural numbers but we use bold symbols to distinguish members of D from ordinary numbers (i. e., $D = \{\mathbf{0}, \mathbf{1}, \mathbf{2}, \ldots\}$);
2. I is a function that maps 0 to $\mathbf{0}$, 1 to $\mathbf{1}$, 2 to $\mathbf{2}$, and so on; and
3. v is a ordinary valuation function defined by cases as follows:
   (a) $v(\neg A) = T$ if and only if $v(A) = F$;
   (b) $v(A \wedge B) = T$ if and only if $v(A) = v(B) = T$;
   (c) $v((\exists x)A(x)) = T$ if and only if there is an interpretation function I' that differs from I at most in the value of $I(x)$, and such that $v(A(x)) = T$ under that interpretation.

Let us add two new predicates to this language: the predicates "small" or $S(n)$ and "large" or $L(n)$. In order to define the interpretation of these predicates, we need to specify the set $\mathbf{M}$ of models. Let S_1 and S_n be two numbers (i. e., $S_1, S_n \in D$) such that $S_1 < S_n$ and for any number S_i such that $S_1 \leq S_i \leq S_n$, there is corresponding model M_i, where

$$\begin{aligned} I_i(S) &= (I_i^+(S), I_i^-(S)) \\ &= (\{\mathbf{0}, \mathbf{1}, \ldots, S_i\}, \{S_{i+1}, S_{i+2}, \ldots\}). \end{aligned}$$

It is not difficult to see that if $i < j$. then $I_i^+(S) \subseteq I_j^+(S)$. Similarly, we can select two numbers L_1 and L_m, $L_1 > L_m$, such that, for any number L_i, such that $L_1 \geq L_i \geq L_m$, there is corresponding model M_i, where

$$\begin{aligned} I_i(L) &= (I_i^+(L), I_i^-(L)) \\ &= (\{L_{i+1}, L_{i+2}, \ldots\}, \{\mathbf{0}, \mathbf{1}, \ldots, L_i\}). \end{aligned}$$

It is not difficult to see that if $i < j$. then $I_i^+(L) \subseteq I_j^+(L)$.

In order to use these predicates, Van Bendegem assumes that $n = m$ because this make it possible to have for a “large” model for every “small” model and vice versa. Now it is possible to prove the following theorem.

Theorem 2.4.1. $(\forall n)(S(n) \vee \neg S(n)) \wedge (\forall n)(L(n) \vee \neg L(n))$ *or, in words, any number is either small (large) or not small (large).*

Now suppose that for any model M_i, $S_i < L_i$ (or equivalently $S_n < L_n$). Then we can easily prove the following.

Theorem 2.4.2. $(\forall n)(L(n) \Rightarrow \neg S(n)) \wedge (\forall n)(S(n) \Rightarrow \neg L(n))$ *or, in words, if a number n is large (small), then it is not small (large).*

Van Bendegem proved some other interesting theorems, but we do not plan to fully present his work here.

2.5 Context and mathematics

In the previous sections, we presented a form of vague mathematics based on the adoption of the first two approaches to vagueness. To the best of our knowledge, the third approach, that is, contextualism, has not been used to define some kind of alternative mathematics. However, it has been used to promote a rather novel idea, that is, that mathematics is actually context depended. This simply means that mathematics is vague. Of course, the vagueness of mathematics is rather special and this section we will briefly describe two similar approaches.

Löwe and Müller [9] introduced the first approach. In their approach, they try to refute the idea that “mathematical knowledge is assumed absolute and undeniably firm.” People believe this because mathematicians *prove* what they claim and so no one can challenge the produced knowledge. However, most proofs published in books and journal are *informal* since they are not proof expressed in mathematically well-defined systems. But what is a *formal* proof? In general, a proof-tree in first-order logic, that is, a *derivation structure* generated by the application of *inference rules* to *premises*, is considered a formal proof. More generally, a formal proof should consist of very small steps that be mechanically checked by a mechanical procedure (i. e., an algorithm). However, most working mathematicians do not produce formal proofs but people working in proof theory try to do so. An informal proof contains gaps and some times look unfinished (e. g., some authors just invite readers to finish proofs...). Also, in many cases, the structure of a proof depends on the medium it appears. For example, a proof for first year university students is usually quite lengthy and contains many (unnecessary?) details, while a proof published in a mathematical journal is usually much harder to follow and sometimes omit details that are obvious to (very) few experts.

In general, the following statement summarizes the standard view of knowledge in mathematics:

> S knows that P if and only if S has available a proof of P.

Although we will not present the various forms of this statement that Löwe and Müller [9] have managed to derive, still it is rather important to say that there are many examples of proper knowledge attributions where the previous statement is just false. For example, we are certain that Euler knew that

$$1+\frac{1}{4}+\frac{1}{9}+\frac{1}{16}+\frac{1}{25}+\cdots=\frac{\pi^2}{6},$$

but Euler did not have a proof of this fact. Roughly, he used finite sums that approximated the final result and using inductive thinking he established that the formula must be true. Another case where the statement above is not true is about the knowledge of many mathematical truths by nonmathematicians (e. g., engineers, everyday people, etc.). These people know certain mathematical truths (e. g., the Pythagorean theorem) but have no idea how it can be proved! Of course, this happens because they only care about the result but still their case falsifies the statement above.

The previous remark can be used in conjunction to David Lewis's [7] context dependent analysis of "S knows that P", which is expressed as follows:

> S knows that P if and only if S's evidence eliminates every possibility in which not-P – Psst! – except for those possibilities that we are properly ignoring.

Without going into the details, Löwe and Müller [9] use the notion of skill to reformulate the previous statement into one that takes into account mathematical knowledge and of course contextualism:

> S knows that P if and only if S's current mathematical skills are sufficient to produce the form of proof or justification for P required by the actual context.

This statement makes clear the connection between mathematical knowledge and contextualism.

According to Liu [8], the contextualistic way to understand mathematics or science is to understand mathematical or scientific contexts, respectively. In addition, the study of mathematical or scientific objects is the study of these objects in context. As regards mathematical context, it consists of three elements: syntax, semantics, and pragmatics. Here, syntax is study of symbols and the ways they can be put together, semantics is about the meaning of symbols, and pragmatics is about the possible connections between a research subject, symbols, and their meaning. Context is the unity of their interactions. This context has boundaries which is determined by a domain of questions. More specifically, if the question is about a sentence, a particular problem, or a particular history, then the boundary of this context is this sentence, this problem, or this historical process, respectively. In case the semantic boundary is exceeded, then the corresponding context will change. In order to make clear this, Liu gives the following example.

Consider the dynamic equations on time scales that follow:

$$y^{\Delta}(t) = f(t, y(t)), \quad t \in T, \tag{1}$$

where $\mathbb{T}$ is a nonempty closed subset of $\mathbb{R}$, f is a continuous function, $y^{\Delta}(t)$ is the Δ derivative of f at t. The definition of the Δ derivative follows.

Definition 2.5.1. A function f on $\mathbb{T}$ is said to be Δ or Hilger differentiable at some point $t \in \mathbb{T}$ if there is a number $f^{\Delta}(t)$ such that for every $\epsilon > 0$ there is a neighborhood $U \subset \mathbb{T}$ such that

$$|f(\sigma(t)) - f(s) - f^{\Delta}(t)(\sigma(t) - s)| \leq \epsilon|\sigma(t) - s|, \quad s \in U,$$

where

$$\sigma(t) = \begin{cases} \inf\{s \in \mathbb{T} \mid t < s\}, & t < \sup \mathbb{T}, \\ \sup \mathbb{T}, & t \geq \sup \mathbb{T}. \end{cases}$$

We say that $f^{\Delta}(t)$ is the Δ or Hilger derivative of f at t.

Suppose that $\mathbb{T} = \mathbb{R}$, then equation 1 reduces to the differential equation

$$y'(t) = f(t, y(t)),$$

that describes a continuous process. If $\mathbb{T} = \mathbb{Z}$, then equation 1 reduces to the difference equation

$$y(t+1) - y(t) = f(t, y(t)),$$

that describes a discrete process. Finally, if $\mathbb{T} = [0, 1] \cup \mathbb{N}$, then equation (1) reduces to dynamic equation on time scales

$$y^{\Delta}(t) = f(t, y(t)),$$

that describe a process with both continuous and discrete characteristics.

Mathematical context is dynamic rather than static. Large changes in context have a deep effect in the meaning of new contexts. New ideas and new approaches force mathematical context to change and, therefore, to continuously get recontextualized. Practically, this means that new context builds from previous while it incorporates new ideas and developments. A simple example of recontextualization is the emergence of the Lebesgue integral from the Riemann integral. To make things simple, we will assume that integration is the process by which we calculate the area under the curve of a function. Roughly, we divide this area into vertical slices of width Δx and add them up. When the slices approach zero in width, then their sum is the Riemann integral of the function. If we divide the area under the curve of a function into horizontal slices whose width approach zero, then their sum is the Lebesgue integral of the function.

2.6 Vagueness and the law of excluded middle

Mathematics is the *language* we use to express ideas and to reason in science and technology. Today most researchers and scholars assume that any mathematical statement is either true or false. This "fact" is known as the *law of excluded middle* (LEM). The law is also known by its Latin name: *tertium non datur*. That LEM is true means that the symbolic expression $A \vee \neg A$ is always true (i. e., it is a tautology). Obviously, the changes associated to each logical approach to vagueness affect the validity of LEM. In fuzzy logic, we assume that a truth value is any real number that belongs to the unit interval, $A \vee B = \max(A, B)$, and $\neg A = 1 - A$. To see what this entails, let us consider the following (partial) truth table:

A	$\neg A \vee A$
0.0	1.0
0.1	0.9
0.2	0.8
0.3	0.7
0.4	0.6
0.5	0.5
0.6	0.6
0.7	0.7
0.8	0.8
0.9	0.9
1.0	1.0

There is no question that LEM is not valid for fuzzy logic. On the other hand, LEM is always true under supervaluationism.[7] Arnon Avron and Beata Konikowska [2] have proposed a three-valued logic. The truth values are **t**, **u**, and **f**. It holds that $\neg\mathbf{t} = \mathbf{f}$, $\neg\mathbf{f} = \mathbf{t}$, and $\neg\mathbf{u} = \mathbf{u}$.

$\vee$	f	u	t
f	f	u	t
u	u	{u, t}	t
t	t	t	t

From these *facts*, we can conclude that LEM does not really hold for this logic of rough sets. In conclusion, we can say LEM is a mixed bag as far it regards the various approaches to vagueness.

7 See Andrea Iacona, *Future Contingents*, Internet Encyclopedia of Philosophy, https://iep.utm.edu/fut-cont/ (last accessed February 27, 2021).

2.7 Conclusions

We have presented various approaches to the creation of vague mathematics. What is really interesting is that all expressions of vagueness can be used to build alternative mathematics. Of course, not all approaches have reached the same degree of maturity but the important thing is that vague mathematics is there and they can be used to describe our vague world.

Bibliography

[1] K. Akiba and A. Abasnezhad, Vague Objects and Vague Identity. Logic, Epistemology, and the Unity of Science, Vol. 33. Springer, Dordrecht, The Netherlands, 2014.
[2] A. Avron and B. Konikowska, Rough Sets and 3-Valued Logics, Stud. Log. 90(1) (2008), 69–92.
[3] C. L. Chang, Fuzzy topological spaces, J. Math. Anal. Appl. 24(1) (1968), 182–190.
[4] W. P. Hanage, Fuzzy species revisited, BMC Biol. 11(1) (2013).
[5] W. P. Hanage, C. Fraser and B. G. Spratt, Fuzzy species among recombinogenic bacteria, BMC Biol. 3(1) (2005).
[6] D. Klaua, Ein Ansatz zur mehrwertigen Mengenlehre, Math. Nachr. 33(5–6) (1967), 273–296.
[7] D. Lewis, Elusive Knowledge, Australas. J. Philos 74(4) (1996), 549–567.
[8] J. Liu, A Contextualist Interpretation of Mathematics, in: Scientific Explanation and Methodology of Science, G. Guo and C. Liu, eds, World Scientific, Singapore, 2014, pp. 128–137.
[9] B. Löwe and T. Müller, Mathematical Knowledge is Context Dependent, Grazer Philos. Stud. 76(1) (2008), 91–107.
[10] C. Mortensen, Inconsistent Mathematics. Mathematics and Its Applications, Vol. 312, Kluwer Academic Publishers, Dordrecht, The Netherlands, 1995.
[11] Z. Pawlak, Rough Sets, Int. J. Comput. Inf. Sci. 11(5) (1982), 341–356.
[12] Z. Pawlak, Vagueness — A rough set view, in: Structures in Logic and Computer Science: A Selection of Essays in Honor of A. Ehrenfeucht, J. Mycielski, G. Rozenberg and A. Salomaa, eds, Lecture Notes in Computer Science, Vol. 1261, Springer, Berlin, Germany, EU, 1997, pp. 106–117.
[13] R. Sorensen, Vagueness, in: The Stanford Encyclopedia of Philosophy, E. N. Zalta, ed., 2018, summer 2018 edn.
[14] A. Syropoulos and T. Grammenos, A Modern Introduction to Fuzzy Mathematics, John Wiley and Sons Ltd, New York, 2020.
[15] J. P. Van Bendegem, Alternative Mathematics: The Vague Way, Synthese 125(1) (2000), 19–31.
[16] J. P. Van Bendegem, Can there be an alternative mathematics, really? in: Activity and Sign: Grounding Mathematics Education, M. H. Hoffmann, J. Lenhard and F. Seeger, eds, Springer US, Boston, MA, 2005, pp. 349–359.
[17] K. Van Deemter, Not Exactly, in: Praise of Vagueness, Oxford University Press, Oxford, UK, 2010.
[18] B. C. Van Fraassen, Singular Terms, Truth-Value Gaps, and Free Logic, J. Philos. 63(17) (1966), 481–495.
[19] B. C. Van Fraassen, Formal Semantics and Logic, Macmillan, New York, 1971, A PDF is available from: https://www.princeton.edu/~fraassen/Formal%20Semantics%20and%20Logic.pdf.
[20] Q. Wu, T. Wang, Y. Huang and J. Li, Topology Theory on Rough Sets, IEEE Trans. Syst. Man Cybern., Part B, Cybern. 38(1) (2008), 68–77.
[21] L. A. Zadeh, Fuzzy Sets, Inf. Control 8 (1965), 338–353.

Apostolos Syropoulos

3 Vague theory of computation

Abstract: Computing is considered by most people as an exact science where nothing can "deviate" from the norm. However, there are many vague problems or ordinary problems that deal with vague data. These cases cannot be easily handled by ordinary algorithms; thus we need tools that can properly handle these problems and these data. Such tools include fuzzy algorithms and fuzzy computing tools, rough set computing tools, and others.

3.1 Introduction

Computing, just like mathematics, is considered an exact and precise science. However, even basic definition in the theory of computation are not that exact and precise. For example, consider the following definition of *algorithm* that is borrowed from [9].

Definition 3.1.1. An algorithm is a finite set of instructions that if followed, accomplish a particular task. In addition, every algorithm must satisfy the following criteria:
(1) *input*: there are zero or more quantities that are externally supplied;
(2) *output*: at least one quantity is produced;
(3) *definiteness*: each instruction must be clear and unambiguous;
(4) *finiteness*: if we trace out the instructions of an algorithm, then for all cases the algorithm will terminate after a finite number of steps;
(5) *effectiveness*: every instruction must be sufficiently basic that it can in principle be carried out by a person using only pencil and paper. It is not enough that each operation be defined as in (3), but it must also be feasible.

Consider finiteness. Suppose a process realizing an algorithm requires many million years to complete. Is it finite? Also, when an algorithm is really effective? How do we know that our approach is the most effective way to compute something? Do we know all the methods, the tools, etc., that can be used to solve a problem? Even if we consider the Turing machine, which is considered the archetypal computing device, then something is considered effective if it can be computed by it. However, programming a Turing machine is quite difficult. Thus many tasks that seem very easy to deal with cannot be easily tackled with the aid of a Turing machine.

What is really interesting is that Turing machines are supposed to be machines that *follow* instruction and never fail. So "it is thus not logically possible for a Turing machine to execute an instruction to 'print' an S_0 and yet fail to 'place' an S_0 on the tape" [4, p. 168]. However, there is no hardware that is not faulty, that is, sooner of later hardware will fail. But failure is something that happens gradually. This means that

https://doi.org/10.1515/9783110704303-003

devices might be faulty to a degree. Clearly, this means that hardware is vague and, therefore, computation is vague. A recent trend in computing is the *approximate computing paradigm* [26]. In this paradigm "hardware and software generate and reason about estimates." In general, approximate computing is about the use of randomness and/or probabilities in programs, still it is quite possible to use vagueness in the form of fuzzy sets in programming languages and computing in general (e. g., see [20, 21] for a brief presentation of such tools).

Plan of the chapter

Since there are four expressions of vagueness (see [20]), we will present how fuzzy sets and rough sets are used in computing, how paraconsistent logics are used in computing, how context affects computing, and how supervaluation is used in computational logic.

3.2 Fuzzy computation

As expected, the founder of fuzzy set theory outlined the idea of fuzzy computation. In particular, Lotfi Aliasker Zadeh [27] proposed that *fuzzy algorithms* should be algorithms that would contain instructions like the following:

(a) *Set y approximately equal to* 10 *if x is approximately equal to* 5, or
(b) *If x is large, increase y by several units*, or
(c) *If x is large, increase y by several units; if x is small, decrease y by several units; otherwise keep y unchanged.*

Obviously, the source of vagueness in these instructions are the fuzzy sets that are identified by their underlined names. In addition, he proposed how these instructions should be executed. His idea is quite dated now so there is no reason to mention it. Also, in the same paper he introduced the idea of a fuzzy Turing machine but his description was rather informal. However, later on many researchers took over the difficult task to precisely define fuzzy Turing machines and of the fuzzy models of computations (see [19] for an overview of the theory of fuzzy computation).

I am convinced that all flourishing fields of research have some philosophical basis. This means that researchers working in the field share the same principles and basic ideas or, else, there could not be no consensus among them. The net effect of such a disparity of ideas could be catastrophic for a given field. Thus it is more than necessary to formulate a philosophy of any field of research. I have already tried to define such a philosophy [23] and I am going to use it to describe the field of fuzzy computation.

First of all, it is necessary to understand what belongs to the field of fuzzy computation. Things like fuzzy arithmetic, fuzzy databases, fuzzy web searches, etc., do not belong to this field. In fact, these are applications of "vague-like" computation. To make this clear, consider a fuzzy database. This database operates in a nonvague environment and is supposed to handle vague data. Obviously, this is quite useful, but it is far from being a vague computation. In different words, a fuzzy database is a form of fuzzy computation if a simulator of a quantum computer than runs on conventional hardware is able to achieve exactly what a real quantum computer can do.

On the other hand, fuzzy computation should include tools able to compute in a vague environment. Such tools could be conceptual computing devices that operate in an environment, where, for instance, hardware is far from being "digital" and commands can be executed to some degree. It is important to understand that I am not proposing a new computing paradigm but a new approach to the essence of computation. Thus it would make sense to question the *limits of computation*. A fuzzy version of the Turing machine is a fuzzy conceptual computing devices. Although a fuzzy version of the Turing machine may seem like an ideal fuzzy conceptual computing device (see [23] for an up-to-date presentation of fuzzy Turing machines), one should note that no one uses quantum Turing machines to study quantum computation. A more natural model of fuzzy computation are fuzzy P systems (see [16, 17] but see also [19]).

The most general form of fuzzy P systems uses the notion of *L-fuzzy multisets* that are characterized by a high-order function, where L is a frame.[1] In particular, an L-fuzzy multiset A is characterized by a function

$$A : X \to \mathbb{N}^L,$$

where $\mathbb{N}$ is the set of natural numbers including zero. It is not difficult to see that any L-fuzzy multiset A is actually characterized by a function

$$A : X \times L \to \mathbb{N},$$

which is obtained from the former function by *uncurrying* it. However, it is more natural to demand that for each element x there is only one membership degree and one multiplicity. In other words, an "L-fuzzy multiset" A should be characterized by a function $X \to L \times \mathbb{N}$. To distinguish these structures from fuzzy multisets, I call them *multifuzzy* sets [16]. Given a multifuzzy set, A, the expression $A(x) = (\ell, n)$ denotes that there are n copies of x that belong to A with degree that is equal to ℓ. The cardinality

1 A poset A is a *frame* if and only if
1. every subset has a join
2. every finite subset has a meet
3. binary meets distribute over joins:

$$x \wedge \bigvee Y = \bigvee \{x \wedge y : y \in Y\}.$$

of an L-fuzzy multiset A is the sum

$$\operatorname{card} A = \sum_{a \in X} A_m(a) A_\mu(a),$$

where $A(a) = (A_\mu(a), A_m(a))$.

P systems is a model of computation inspired by the way cells live and function. The model was introduced by Gheorghe Păun [12]. The model builds upon nested compartments surrounded by porous *membranes*. It is quite instructive to think of the membrane structure as a bubbles-inside-bubbles structure, that is, the top bubble may contain bubbles, which, in turn, may contain other bubbles, etc. Initially, each compartment contains a number of possible repeated objects (i. e., a multiset of objects). In addition, each compartment is associated with a number of rules that dictate what should happen to the elements contained in the compartment. In the simplest case, these rules are just multiset rewriting rules. Once "computation" commences, the compartments exchange objects according to the rules. The computation stops once no rule can be applied. The result of the computation is equal to the number of objects (i. e., the cardinality of the multiset) that reside in a designated compartment that is called the *output membrane*. In order to give a formal definition of fuzzy P systems, we need an auxiliary definition:

Definition 3.2.1. Let $V = \{[,]\}$ be an alphabet. Then the set MS is the least set inductively defined as follows:
1. $[] \in \mathrm{MS}$;
2. if $\mu_1, \mu_2, \ldots \mu_n \in \mathrm{MS}$, then $[\mu_1 \ldots \mu_n] \in \mathrm{MS}$.

Now we are ready to give the formal definition of fuzzy P systems.

Definition 3.2.2. A P system with fuzzy data is a construction

$$\Pi_{\mathrm{FD}} = (O, \mu, w^{(1)}, \ldots, w^{(m)}, R_1, \ldots, R_m, i_0),$$

where:
1. O is an alphabet (i. e., a set of distinct entities) whose elements are called *objects*;
2. μ is the membrane structure of degree $m \geq 1$, which is the depth of the corresponding tree structure; membranes are injectively labeled with successive natural numbers starting with one;
3. each $w^{(i)} : O \to L \times \mathbb{N}$, $1 \leq i \leq m$ is a multifuzzy set over O associated with the region surrounded by membrane i;
4. R_i, $1 \leq i \leq m$, are finite sets of multiset rewriting rules (called *evolution rules*) over O. An evolution rule is of the form $u \to v$, $u \in O^*$ and $v \in O_{\mathrm{TAR}}^*$, where $O_{\mathrm{TAR}} = O \times \mathrm{TAR}$,

$$\mathrm{TAR} = \{\mathrm{here}, \mathrm{out}\} \cup \{\mathrm{in}_j \mid 1 \leq j \leq m\}.$$

The keywords "here," "out," and "in$_j$" are used to specify the current compartment (i. e., the compartment the rule is associated with), the compartments that surrounds the current compartment, and the compartment with label j, respectively. The effect of each rule is the removal of the elements of the left-hand side of the rule from the current compartment (i. e., elements that match the left-hand side of a rule are removed from the current compartment) and the introduction of the elements of the right-hand side to the designated compartments. Also, the rules implicitly transfer the fuzzy degrees to membership in their new "home set";
5. $i_0 \in \{1, 2, \ldots, m\}$ is the label of an elementary membrane (i. e., a membrane that does not contain any other membrane), called the *output* membrane.

But we can go one step further. Thus instead of having crisp evolution rules, we can have fuzzy evolution rules. These rules have the following general form:

$$u \xrightarrow{\lambda} u$$

and they are associated with a plausibility degree, λ, that indicates to which degree the current rule is applicable. Obviously, this reminds of a stochastic process but here there is no randomness involved in choosing a rule. The most plausible rules are always selected. Also, the plausibility degree modifies the membership degree of the elements of the multifuzzy set on which a rule is applied. In particular, if any "molecule" produced by an evolution rule whose has membership degree $\lambda_1 > \lambda$, where λ is the plausibility degree of the fuzzy evolution rule, then its membership degree will become λ.

Now that I have introduced fuzzy evolution rules, I can introduce fuzzy labeled transition systems.

Definition 3.2.3. A fuzzy labeled transition system over a crisp set of actions $\mathscr{A}$ is a triple $(\mathscr{A}, \mathscr{Q}, \mathscr{T})$ consisting of
- a set $\mathscr{Q}$ of states; and
- a ternary fuzzy relation $\mathscr{T}(\mathscr{Q}, \mathscr{A}, \mathscr{Q})$ called the *fuzzy transition relation*.

If the membership degree of (q, α, q') is d, then $q \underset{d}{\overset{\alpha}{\rightarrow}} q'$ denotes that the plausibility degree to go from state q to state q' by action α is d. More generally, if $q_0 \underset{d_1}{\overset{\alpha_1}{\rightarrow}} q_1 \underset{d_2}{\overset{\alpha_2}{\rightarrow}} q_2 \cdots \underset{d_n}{\overset{\alpha_n}{\rightarrow}} q_n$, then q_n is called a *derivative* of q_0 with plausibility degree equal to $\min\{d_1, d_2, \ldots, d_n\}$.

Of course, one can say much more about fuzzy labeled transition systems, but my purpose here is to show how we can define fuzzy models of computation and not to study all the properties of such systems. The reader can find more about such models of computation in [19]. In addition, the idea of vague computers, in general, and an "interpretation" of quantum computers as vague computers, in particular, have been also explored (see [18, 22]).

3.3 Rough set computation

In this section, I present a rough sketch of a rough Turing machine [19]. In a nutshell, a rough Turing machine should compute a result by following the general idea of rough sets, that is, it should compute an upper and a lower approximation of the result. In order to be able to compute the upper and the lower approximations, a rough Turing machine should be actually a pair of Turing machines that differ only in their controlling devices. Initially, both tapes should have the same data printed on them. The machine halts when action on both tapes stops.

It is a fact that the transition function of a nondeterministic Turing machine returns a subset of $Q \times \Gamma \times \{L, R, N\}$. In addition, the transition function of an ordinary automaton returns a state, the transition function of a nondeterministic automaton returns a set of states, and the transition function of a rough automaton returns a rough set of states. Therefore, one should expect that the transition function of a rough Turing machine returns a rough set. The following is a first attempt to formally define a rough Turing machine based on the previous remarks.

Definition 3.3.1. A rough Turing machine $\mathscr{M}$ is an octuple $(Q, \Sigma, \Gamma, \delta, \sqcup, \triangleright, q_0, H, R)$, where

- R is an equivalence relation on $Q \times \Gamma \times \{L, R, N\}$;
- δ is a function from $((Q \setminus H) \times \Gamma) \times ((Q \setminus H) \times \Gamma)$ to $\mathbf{A} \times \mathbf{A}$, where

$$\mathbf{A} = \{(\underline{A}, \overline{A}) \mid A \subseteq Q \times \Gamma \times \{L, R, N\}\}$$

  is the set of all definable sets in the approximation space $(Q \times \Gamma \times \{L, R, N\}, R)$, such that $\delta(q_i, s_j, q_k, s_l) = (\underline{A}, \overline{A})$ is a rough set that belongs to $\mathbf{A}$;
- and all others are as in the case of an ordinary Turing machine.

3.4 Paraconsistent logics and computation

Following Manuel Bremer [3], we will assume that a logic is a proof system and its corresponding semantics. When the difference between syntax and semantics does matter, then the proof system is the systematisation of the relation of derivability, $\vdash$, whereas the semantic consequence relation, $\models$, is used for semantic inferences.

Definition 3.4.1. A logic is *paraconsistent* if and only if a set of statements can be simply or explicitly inconsistent without being trivial with respect to the relations $\vdash$ and $\models$ as defined by the logic. In particular, this means that

$$(\exists A)(A \in \Gamma \wedge \neg A \in \Gamma) \wedge (\exists B)\Gamma \nvdash B.$$

In simple words, a logic is paraconsistent if it contains inconsistencies, which means that both A and $\neg A$ are provable. However, not all pairs of such statements are provable or else we end up with a trivial logic where everything is true.

If we are not allowed to say that something has a property to some degree, then we can only say that it has and it does not have this property. Thus if my t-shirt is blue to some degree, then it is both blue and not blue. This implies that paraconsistent logics can be used to model vagueness. Indeed, modern pioneers of paraconsistent logics argued in favor of the use of them in vagueness. For example, Stanisław Jaśkowski [10] discussed this idea, while, more recently, Zach Weber [24] explained how it is possible to give an interpretation of the sorites paradox in paracosnistent logics. Of course, here I am not interested in vagueness in general but in vagueness in computation. Thus the first question is: Does it make sense to talk about paraconsistent computation?

Weber [25] presented a very simple yet elegant argument that shows that it does make sense to talk about paraconsistent computation. Suppose that $\mathbb{A}$ is the list of algorithmic functions in just one variable. Also, suppose that F_i is the $(i+1)$th member of $\mathbb{A}$, and f_i is the corresponding function. Define

$$\eth(x) := f_x(x) + 1$$

In order to compute $\eth(x)$, we have to find the $(x + 1)$th element of $\mathbb{A}$, then we have to compute $f_x(x)$ and add one. The steps required to implement this process satisfy every definition of an algorithm, therefore the computation of $\eth$ is an algorithm. Since $\eth$ is a function computed by an algorithm, there is a z such that $\eth = f_z$ corresponding to some F_z on $\mathbb{A}$. This means that

$$f_z(z) = \eth(z) = f_z(z) + 1$$

and so $n = n + 1$ for some natural number n. This means that $\eth$ is an inconsistent computation.

A Turing machine consists of a tape that is divided into read/write cells, a scanning head that can read from cells and write to cells. The behavior of a Turing machine is determined from a set of quadruples and the symbols that are initially printed on its tape. There are four kinds of quadruples:

(i) $q_i S_j S_k q_l$.
(ii) $q_i S_j R q_l$.
(iii) $q_i S_j L q_l$.
(iv) $q_i S_j q_k q_l$.

The letters q_n and S_m represent *states* of the machine and the symbol that is printed on the cell that the scanning head reads, respectively. Thus a quadruple specifies that if the scanning head has just read the symbol S_j and the machine is in state q_i, then it should print the third symbol of the quadruple and enter the state specified by the fourth element of the quadruple. If the third symbol is either the letter R or the letter L, then scanning head will move one cell to the right or left, respectively. The fourth quadruple is used only in so-called oracle Turing machines. In this case, when the machine is in state q_i and the scanning head has read the symbol S_j, then an external operator/agent will decide whether the machine will enter the state q_k or the state q_l. Now a Turing machine is defined to be a finite and nonempty set of quadruples that does not contain two quadruples whose first two symbols are the same [5]. We can call this the consistency restriction. Obviously, by defining a Turing machine that includes at least two quadruples whose first two symbols are the same, we actually define a inconsistent Turing machine.

Paraconsistent Turing machines (*ParTM*) have been first discussed in [2] and their theory was elaborated in [1]. Actually, a ParTM is a nondeterministic Turing machine, that is, a Turing machine that at each step can randomly choose from a set of quadruples.

Definition 3.4.2. A ParTM is a nondeterministic Turing machine such that:
(i) When the machine reaches an ambiguous configuration, it simultaneously executes all possible instructions, which can produce a multiplicity on states, positions, and symbols in some cells of the tape;
(ii) Each instruction is executed in the position corresponding to the respective state; symbols in cells unmodified by the instructions are carried to the next instant of time;
(iii) Inconsistency (or multiplicity) conditions are allowed on the first two symbols of the instructions (as described above); and
(iv) The machine stops when there are no instructions to be executed; at this stage some cells of the tape can contain multiple symbols, any choice of them represents a result of the computation.

The discussion above allows us to define a paraconsistent variant of P systems. The evolution rules of such a system can be of the form $u \to v_1, u \to v_2, \ldots, u \to v_n$, and when more than one of these rules is applicable, then all these rules are applied simultaneously. As in the "normal" case, the system stops when no rule is applicable and the result of the computation is equal to number of elements that a specific compartment contains. It is even possible to define a fuzzy paraconsistent version of P systems. In this case, the rules are associated with a plausibility degree and when the machine can apply a set of evolution rules at the same time, then the plausibility degree is the one that dictates the order in which the rules will be applied. Naturally, if two or more rules have the same plausibility degree, then they are applied simultaneously.

3.5 Context-aware computing

Context-aware computing [7, 13] is not explicitly related to vagueness, at least this is what most people working in the field assume. However, intuitively speaking, there should be some relationship since vagueness *depends* on context. But what do we mean by context, at least from a computational point of view? Many authors consider the definition given by Dey, Abowd, and Salber [6] as the best one. Here is their definition:

Context Any information that can be used to characterize the situation of entities (i. e., whether a person, place, or object) that are considered relevant to the interaction between a user and an application, including the user and the application themselves. Context is typically the location, identity, and state of people, groups, and computational and physical objects.

It is obvious that the authors tried to give a rigorous definition but there is no question that this definition is vague. So is context something vague? Yes! It is and Shapiro [15] has pointed out that context is vague and is itself context sensitive. In addition, he proposed that the technical notion of "partial interpretation" is a rough analog of context, with emphasis on the word "rough." Since context is vague, this implies that context-aware computing is actually some sort of vague computing.

Virtual assistants are the best example of context-aware computing. For example, Amazon's Alexa is a voice-controlled virtual assistant. Alexa was inspired by the conversational computer onboard Star Trek's Starship Enterprise. Alexa can do many things. For example, it can play music, check the weather, control one's smart home, answer questions, do math, tell stories, tell jokes, order dinner, and shop on Amazon. Naturally, many of these activities depend of the context. Thus shopping at Amazon or ordering dinner depends on the shopping habits of the owner and her nutritional needs, respectively. Of course, shopping habits may change whereas nutritional needs depend on many different and unrelated factors (e. g., the weather and one's mood). In order to better tackle these human characteristics, some systems rely on techniques that employ fuzzy sets and fuzzy logic [14]. Since context-aware computing is a very broad field that is still under development, I will stop here my short discussion.

3.6 Supervaluation and computing

To the best of my knowledge, supervaluation has not been used in programming. Most computing systems use the bivalent logic and a growing number of systems, programming languages, etc., indirectly or directly use some fuzzy logic. Thus if one could "merge" supervaluation with a fuzzy logic, then the resulting logic could be used in computing. Although the two approaches to vagueness seem to be incompatible, still

Christian G. Fermüller and Robert Kosik [8] have shown that it is possible to define a logic that combines supervaluationism and fuzzy logics. In particular, their logic, which is called **SŁ**, extends Łukasiewicz logic, but incorporates also classical logic. In this language, formulas consist of propositional variables $p \in V = \{p_1, p_2, \ldots\}$ and the constant $\bot$ (falsum) using the connectives and (strong conjunction) and $\supset$ (implication). These are things provided also by the Łukasiewicz logic. In Łukasiewicz logic, a valuation of formulas involving these connectives is using the Łukasiewicz t-norm $x * y = \sup\{0, x+y-1\}$ and the associated residuum $x \Rightarrow y = \inf\{1, 1-x+y\}$. Typically, we use valuation functions to give to each propositional variable a truth value. For the Łukasiewicz logic, a valuation function is a function $v : V \to [0,1]$ that is extended to v^* for formulas by

$$v^*(A \,\&\, B) = v^*(A) * v^*(B), \quad v^*(A \supset B) = v^*(A) \Rightarrow v^*(B), \quad \text{and} \quad v^*(\bot) = 0.$$

Roughly, a formula F is valid if and only if $v^*(F) = 1$.

The logic **SŁ** has the previous connectives and the connectives $\neg$ (negation), $\wedge$ (weak conjunction), and $\vee$ (disjunction) are defined as follows: $\neg A = A \supset \bot$, $A \wedge B = A \,\&\, (A \supset B)$, and $A \vee B = ((A \supset B) \supset B) \wedge ((B \supset A) \supset A)$. In addition, it included the connective S "pronounced" *it is supertrue that...* Also, this logic uses a precisification space instead of a valuation function to give the semantics of formulas. A precisification space is a triple $\langle W, e, \mu\rangle$, where $W = \{\pi_1, \pi_2, \ldots\}$ is a nonempty set and each π_i is called a precisification point, e is a mapping $W \times V \to \{0,1\}$, and μ is a probability measure on the σ-algebra formed by all subsets of W. For a precisification space $\Pi = \langle W, e, \mu\rangle$, the local truth value $\|A\|_\pi$ is defined for every formula A and every precisification point $\pi \in W$ inductively by

$$\begin{aligned}
\|p\|_\pi &= e(\pi, p), \quad \text{for } p \in V\\
\|\bot\|_\pi &= 0\\
\|A \,\&\, B\|_\pi &= \begin{cases} 1, & \text{if } \|A\|_\pi = 1 \text{ and } \|B\|_\pi = 1\\ 0, & \text{otherwise}\end{cases}\\
\|A \supset B\|_\pi &= \begin{cases} 1, & \text{if } \|A\|_\pi = 1 \text{ and } \|B\|_\pi = 0\\ 0, & \text{otherwise}\end{cases}\\
\|\mathsf{S}A\|_\pi &= \begin{cases} 1, & \text{if } \forall \sigma \in W : \|A\|_\pi = 1\\ 0, & \text{otherwise}\end{cases}
\end{aligned}$$

Local truth values are classical and do not depend on the t-norm we use; nevertheless, the global truth value $\|A\|_\Pi$ of a formula A is defined as follows:

$$\begin{aligned}
\|p\|_\Pi &= \mu(\{\pi \in W \mid e(\pi, p) = 1\}), \quad \text{for } p \in V\\
\|\bot\|_\Pi &= 0
\end{aligned}$$

$$\|A \mathbin{\&} B\|_\Pi = \|A\|_\Pi * \|B\|_\Pi$$
$$\|A \supset B\|_\Pi = \|A\|_\Pi \Rightarrow \|B\|_\Pi$$
$$\|\mathsf{S}A\|_\Pi = \|\mathsf{S}A\|_\pi \quad \text{for any } \pi \in W$$

It should be clear that $\|\mathsf{S}A\|_\pi$ has the same value (either 0 or 1) for all $\pi \in W$. This means that the “local” supertruth is in fact global; which justifies the above clause for $\|\mathsf{S}A\|_\Pi$.

A formula A is valid in SŁ if $\|\mathsf{S}A\|_\Pi = 1$ for all precisification spaces Π.

Proposition 3.6.1. *When the logical system SŁ is restricted to formulas that do not include the operator* S*, the logical system is actually Łukasiewicz logic. On the other hand,* $\{A \mid \mathsf{S}A \in S\text{Ł}\}$ *is actually the modal logic S5 of Clarence Irving Lewis and Cooper Harold Langford* [11].

A Hilbert-style calculus for SŁ can be obtained by extending any system for Ł with the following axioms:

A1 $\mathsf{S}(A \vee \neg A)$;
A2 $\mathsf{S}A \vee \neg \mathsf{S}A$;
A3 $\mathsf{S}(A \supset B) \supset (\mathsf{S}A \supset \mathsf{S}B)$;
A4 $\mathsf{S}A \supset A$;
A5 $\mathsf{S}A \supset \mathsf{S}\mathsf{S}A$;
A6 $\neg\mathsf{S}A \supset \mathsf{S}\neg\mathsf{S}A$;
A7 $\frac{A}{\mathsf{S}A}$.

The last axioms are called the necessitation rule for supertruth.

3.7 Conclusions

In this chapter, I gave a brief overview of various approaches to vague computing. My aim was not to give the full picture but to show that vague computing is something real. In addition, the discussion has provided evidence that vague computing can handle better than ordinary computing a big number of problems, in particular problems that deal with vague data and vague situations. I am strongly convinced that there is a bright future for vague computing in general.

Bibliography

[1] J. C. Agudelo and W. Carnielli, Paraconsistent Machines and their Relation to Quantum Computing, J. Log. Comput. 20(2) (2009), 573–595. Also available as arXiv:0802.0150v2 (quant-ph).

[2] J. C. Agudelo and A. Sicard, Máquinas de Turing paraconsistentes: una posible definición, Matemáticas: Enseñanza Universitaria XII 2 (2004), 37–51. Available at: http://revistaerm.univalle.edu.co/Enlaces/volXII2.html.
[3] M. Bremer, An Introduction to Paraconsistent Logics. Peter Lang, Frankfurt am Main, 2005.
[4] C. E. Cleland, On Effective Procedures, Minds Mach. 12 (2002), 159–179.
[5] M. Davis, Computability and Unsolvability, Dover Publications, Inc., New York, 1982.
[6] A. K. Dey, G. D. Abowd and D. Salber, A Conceptual Framework and a Toolkit for Supporting the Rapid Prototyping of Context-Aware Applications, Hum.-Comput. Interact. 16(2–4) (2001), 97–166.
[7] L. Feng and Context-Aware Computing. Advances in Computer Science, Vol. 3, De Gruyter, Berlin, 2017.
[8] C. G. Fermüller and R. Kosik, Combining Supervaluation and Degree Based Reasoning Under Vagueness, in: Logic for Programming, Artificial Intelligence, and Reasoning, M. Hermann and A. Voronkov, eds, Springer, Berlin, 2006, pp. 212–226.
[9] E. Horowitz and S. Sahni, Fundamentals of Data Structures in Pascal, Computer Science Press, 1984.
[10] S. Jaśkowski, Propositional calculus for contradictory deductive systems, Stud. Soc. Sci. Torun., Sect. A, Stud. Log. 24(1) (1969), 143–157. Originally appeared in Polish under the title Rachunek zdań dla systemów dedukcyjnych sprzecznych in: Studia Societatis Scientiarum Torunensis, Sectio A, Vol. I, No. 5, Toruń 1948.
[11] C. I. L. Lewis and C. H. Langford, Symbolic Logic, 2nd edn, Dover Publications, New York, 1959.
[12] G. Păun, Computing with Membranes, J. Comput. Syst. Sci. 61(1) (2000), 108–143.
[13] R. Porzel, Contextual Computing: Models and Applications, Springer-Verlag, Berlin, 2011.
[14] X. Rong, A. Fourney, R. N. Brewer, M. R. Morris and P. N. Bennett, Managing uncertainty in time expressions for virtual assistants, in: Proceedings of the 2017 CHI Conference on Human Factors in Computing Systems, CHI '17, Association for Computing Machinery, New York, NY, USA, 2017, pp. 568–579.
[15] S. Shapiro, Vagueness in Context, Clarendon Press, Oxford, UK, 2006.
[16] A. Syropoulos, Fuzzifying P Systems, Comput. J. 49(5) (2006), 619–628.
[17] A. Syropoulos, On Generalized Fuzzy Multisets and their Use in Computation, Iran. J. Fuzzy Syst. 9(2) (2012), 115–127.
[18] A. Syropoulos, Vague Computing Is the Natural Way to Compute! in: On Fuzziness: A Homage to Lotfi A. Zadeh – Volume 2, R. Seising, E. Trillas, C. Moraga and S. Termini, eds, Springer, Berlin, 2013, pp. 653–658.
[19] A. Syropoulos, Theory of Fuzzy Computation. IFSR International Series on Systems Science and Engineering, Vol. 31. Springer-Verlag, New York, 2014.
[20] A. Syropoulos, Fuzzy Multisets and Fuzzy Computing, in: Handbook of Research on Generalized and Hybrid Set Structures and Applications for Soft Computing, S. J. John, ed., IGI Global, Hershey, PA, USA, 2016, pp. 23–42.
[21] A. Syropoulos, Fuzzy Sets and Fuzzy Logic, in: Encyclopedia of Computer Science and Technology, P. A. Laplante, ed., CRC Press, Boca Raton, FL, USA, 2016, pp. 459–468.
[22] A. Syropoulos, On Vague Computers, in: Emergent Computation: A Festschrift for Selim, G. Akl and A. Adamatzky, eds, Springer International Publishing, Cham, Switzerland, 2017, pp. 393–402.
[23] A. Syropoulos, A (Basis for a) Philosophy of a Theory of Fuzzy Computation, Kairos. J. Philos. Sci. 20(1) (2018), 181–201.
[24] Z. Weber, A paraconsistent model of vagueness, Mind 119(476) (2010), 1025–1045.
[25] Z. Weber, Paraconsistent Computation and Dialetheic Machines, in: Logical Studies of Paraconsistent Reasoning in Science and Mathematics, H. Andreas and P. Verdée, eds, Springer International Publishing, Cham, Switzerland, 2016, pp. 205–221.
[26] Q. Xu, T. Mytkowicz and N. S. Kim, Approximate Computing: A Survey, IEEE Des. Test 33(1) (2016), 8–22.
[27] L. A. Zadeh, Fuzzy Algorithms, Inf. Control 12 (1968), 94–102.

Nikos Mylonas and Basil Papadopoulos

4 Fuzzy hypotheses tests with crisp data using nonasymptotic fuzzy estimators

Abstract: In classical crisp statistics, the problem of testing a hypothesis is to decide whether to reject or not a hypothesis using a test statistic and either critical values or alternatively a p-value. In a fuzzy test, a hypothesis is rejected or not with a rejection or acceptance degree using fuzzy ordering between a fuzzy test statistic produced by fuzzy estimators and fuzzy critical values. An alternative way to carry out such a test is to use ordering between the fuzzy significance level and the fuzzy p-value. This approach is particularly useful in critical situations, where subtle comparisons between almost equal statistical quantities have to be made. In such cases, the fuzzy hypotheses tests give better results than the crisp ones, since they give us the possibility of partial rejection or acceptance of H_0 using a rejection or acceptance degree.

4.1 Introduction

In classical crisp statistics, the problem of testing a hypothesis for a parameter θ of the distribution of a random variable X is to decide whether to reject or accept the null hypothesis $H_0 : \theta = \theta_0$ at a significance level γ against the alternative hypothesis H_1 from a random sample of observations of X, using a test statistic U, which is evaluated for the sample observations, resulting in a value u. The space of possible values of U is decomposed into a rejection region and its complement, the acceptance region. Depending on the alternative hypothesis H_1, the rejection region has one of the following forms:

(a) $U \leq u_{c1}$, if the alternative hypothesis is $H_1 : \theta < \theta_0$ (one-sided test from the left), where

$$P(U \leq u_{c1}) = \gamma \tag{1}$$

(b) $U \geq u_{c2}$, if the alternative hypothesis is $H_1 : \theta > \theta_0$ (one-sided test from the right), where

$$P(U \geq u_{c2}) = \gamma \tag{2}$$

(c) $U \leq u_{c3}$ or $U \geq u_{c4}$, if the alternative hypothesis is $H_1 : \theta \neq \theta_0$ (two-sided test), where

$$P(U \leq u_{c3}) = \frac{\gamma}{2} \quad \text{and} \quad P(U \geq u_{c4}) = \frac{\gamma}{2} \tag{3}$$

and u_{ci} the critical values of the test, which are determined by the distribution of U. So, H_0 is rejected if the value u of the test statistic U is in the rejection region and not rejected if u in the acceptance region.

https://doi.org/10.1515/9783110704303-004

An equivalent way to test the null hypothesis H_0 against the alternative hypothesis H_1 is to calculate the p-value which is defined for the above cases (a), (b), and (c) as (u is the value of the test statistic for the given sample)

$$\text{(a)} \quad p = P(U \leq u), \quad \text{(b)} \quad p = P(U \geq u), \quad \text{(c)} \quad p = 2\min\{P(U \leq u), P(U \geq u)\} \tag{4}$$

If the p-value is less than γ, then H_0 is rejected at the significance level γ, otherwise H_0 is not rejected.

Several approaches have been developed for fuzzy hypotheses testing, as in [8, 15, 17, 18, 19] where fuzzy hypotheses are tested using crisp data in [15] or fuzzy data in [8]. Among them, one has been proposed in [2, 4, 5] for testing crisp hypotheses from crisp data, which uses fuzzy critical values and fuzzy test statistics constructed by fuzzy estimators, which are produced by a set of confidence intervals.

The nonasymptotic fuzzy estimators developed in [16] generalize the fuzzy estimators of [5]. The former are functions, whereas the latter are not functions, since they contain two small vertical segments.

In [12, 13], fuzzy hypotheses tests which use test statistics constructed by nonasymptotic fuzzy estimators and fuzzy critical values have been developed and in [14] fuzzy hypotheses tests, which use fuzzy p-value produced by these statistics.

In Section 4.2, we introduce the concept of fuzzy estimation of a parameter of the distribution of a random variable X using a sample of observations of X and in the following subsections we develop nonasymptotic fuzzy estimators of:

(1) the mean of a random variable which follows normal distribution with known variance,
(2) the mean of a random variable which follows any distribution from a large sample using the central limit theorem,
(3) the mean of a random variable which follows normal distribution with unknown variance from a small sample,
(4) the difference of the means of two normal random variables with known variances,
(5) the difference of the means of two random variables, which follow any distribution from two large samples,
(6) the difference of the means of two normal random variables with unknown variances from two small samples,
(7) the probability of success of a binomial distribution from a large sample,
(8) the probabilities of success of two binomial distributions from two large samples,
(9) the variance of a normal random variable,
(10) the ratio of the variances of two normal random variables.

In Section 4.3, we describe fuzzy number ordering, since it is necessary in fuzzy hypotheses testing. In Section 4.4, we describe fuzzy hypotheses tests using fuzzy test statistics and fuzzy critical values in Section 4.4.1 or fuzzy p-value and fuzzy significance level in Section 4.4.2. In Sections 4.5–4.10, we apply these two methods in several cases of fuzzy hypotheses tests.

4.2 Fuzzy estimators

The estimation of a parameter of the probability density function (or probability function of a discrete random variable) of a distribution is one of the main purposes of inferential statistics.

Let X be a random variable with probability density function (or probability function for a discrete random variable) $f(x;\theta)$, where θ an unknown parameter which has to be estimated from a random sample $X_l, X_2, \ldots, X_n$ of observations of X. For the estimation of θ, we use a statistic U which is a function of $X_l, X_2, \ldots, X_n$. For a specific sample for which these random variables have values $X_i = x_i, i = 1, 2, \ldots, n$, we can obtain a point estimate $\theta_0 = U(x_i)$ for θ, which is not of high interest, since the probability this to be the required value of θ is zero. A way to estimate θ in crisp statistics is to find a $(1-\beta)100\,\%$ confidence interval for θ in which the value of θ can be found with probability $1-\beta$. Usually $\beta = 0.005$ or 0.01 or 0.02 or 0.05 or 0.1 for $99.5\,\%$, $99\,\%$, $98\,\%$, $95\,\%$, or $90\,\%$ confidence intervals.

Among many other methods a fuzzy estimation method has been proposed in [5], according to which placing all the confidence intervals one on the top of the other a triangular shaped fuzzy number $\overline{\theta}$ is constructed, the α-cuts of which are the confidence intervals starting with $\beta = 0.01$ (we could begin with 0.001 or 0.005, etc.). To finish the "bottom" of $\overline{\theta}$ in order to make it a complete fuzzy number, one needs to drop its graph straight down to complete its α-cuts. So, a fuzzy estimator $\overline{\theta}$ of θ is produced from a given sample with α-cuts

$$\overline{\theta}_B[\alpha] = \begin{cases} [\theta_l(a), \theta_r(a)], & 0.01 \le \alpha \le 1 \\ [\theta_l(0.01), \theta_r(0.01)], & 0 \le \alpha < 0.01 \end{cases} \tag{5}$$

where $[\theta_l(a), \theta_r(a)]$ the $(1-\alpha)100\,\%$ confidence interval of $\overline{\theta}$ deduced from the specific sample. The fuzzy estimator $\overline{\theta}$ produced in this way contains much more information than just a single interval estimate.

The fuzzy estimator $\overline{\theta}$ of (5) is not a function, since it contains two small vertical segments (see Figure 1). Using the linear onto function,

$$h : [0,1] \to \left[\frac{y_0}{2}, \frac{1}{2}\right] \quad h(\alpha) = \left(\frac{1}{2} - \frac{y_0}{2}\right)\alpha + \frac{y_0}{2}, \quad y_0 = 0.01 \tag{6}$$

the fuzzy estimator (5) is generalized in [16] to the nonasymptotic fuzzy estimator (see Figure 2) with α-cuts

$$\overline{\theta}[\alpha] = [\theta_l(2h(a)), \theta_r(2h(a))], \quad 0 \le \alpha \le 1 \tag{7}$$

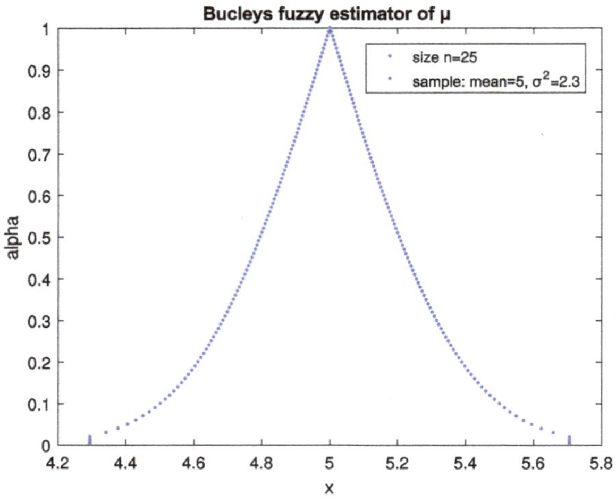

Figure 1: Bucley's fuzzy estimator of μ.

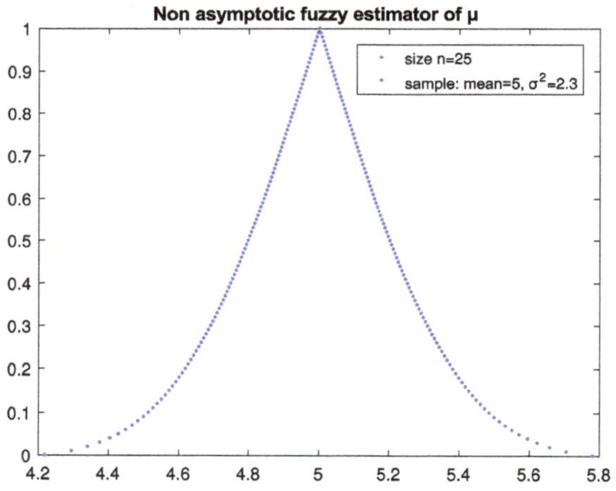

Figure 2: Nonasymptotic fuzzy estimator of μ, known variance.

4.2.1 Estimation of the mean of a normal variable with known variance

If the random variable X follows normal distribution with known variance σ, then the $(1-\beta)100\,\%$ confidence interval of the mean μ of X derived from a random sample of observations of X of size n and sample mean $\overline{x}$ is [9],

$$\left[\overline{x} - z_{\frac{\beta}{2}} \frac{\sigma}{\sqrt{n}}, \overline{x} + z_{\frac{\beta}{2}} \frac{\sigma}{\sqrt{n}}\right] \tag{8}$$

where

$$z_{\frac{\beta}{2}} = \Phi^{-1}\left(1 - \frac{\beta}{2}\right) \tag{9}$$

and Φ^{-1} the inverse distribution function of the standard normal distribution. So, according to (5) the Bucley's fuzzy estimator of μ is [5],

$$\overline{\mu}_B[\alpha] = \begin{cases} [\overline{x} - z_{\frac{\alpha}{2}} \frac{\sigma}{\sqrt{n}}, \overline{x} + z_{\frac{\alpha}{2}} \frac{\sigma}{\sqrt{n}}], & 0.01 \le \alpha \le 1 \\ [\overline{x} - z_{0.005} \frac{\sigma}{\sqrt{n}}, \overline{x} + z_{0.005} \frac{\sigma}{\sqrt{n}}], & 0 \le \alpha < 0.01 \end{cases} \tag{10}$$

and according to (7) the α-cuts of the nonasymptotic fuzzy estimator of μ are [16]

$$\overline{\mu}[\alpha] = \left[\overline{x} - z_{h(\alpha)} \frac{\sigma}{\sqrt{n}}, \overline{x} + z_{h(\alpha)} \frac{\sigma}{\sqrt{n}}\right], \quad 0 \le \alpha \le 1 \tag{11}$$

where $h(\alpha)$ is given by (6) and

$$z_{h(\alpha)} = \Phi^{-1}(1 - h(\alpha)) \tag{12}$$

Example 1. Implementing (10) and (11) for a sample of $n = 25$ observations with sample mean $\overline{x} = 5$ of a random variable, which follows normal distribution with variance $\sigma^2 = 2.3$, we obtain:

in Figure 1, the Bucley's fuzzy estimator of the mean μ of X;
in Figure 2, the nonasymptotic fuzzy estimator of μ.

4.2.2 Estimation of the mean of a random variable from a large sample

If the random variable X follows any distribution, then according to the central limit theorem the $(1-\beta)100\,\%$ confidence interval of the mean μ of X derived from a random sample of observations of X of large size n ($n > 30$) and sample mean and variance $\overline{x}$ and s^2 is [9],

$$\left[\overline{x} - z_{\frac{\beta}{2}} \frac{s}{\sqrt{n}}, \overline{x} + z_{\frac{\beta}{2}} \frac{s}{\sqrt{n}}\right] \tag{13}$$

So according to (7), the α-cuts of the nonasymptotic fuzzy estimator of μ are

$$\overline{\mu}[\alpha] = \left[\overline{x} - z_{h(\alpha)} \frac{s}{\sqrt{n}}, \overline{x} + z_{h(\alpha)} \frac{s}{\sqrt{n}}\right], \quad 0 \le \alpha \le 1 \tag{14}$$

where $z_{h(\alpha)}$ are given by (12) for $h(\alpha)$ of (6).

Example 2. Implementing (14), we obtain in Figure 3 the nonasymptotic fuzzy estimator of μ of a random variable X from a large sample of $n = 100$ observations with sample mean and variance $\overline{x} = 5$ and $s^2 = 1.8$.

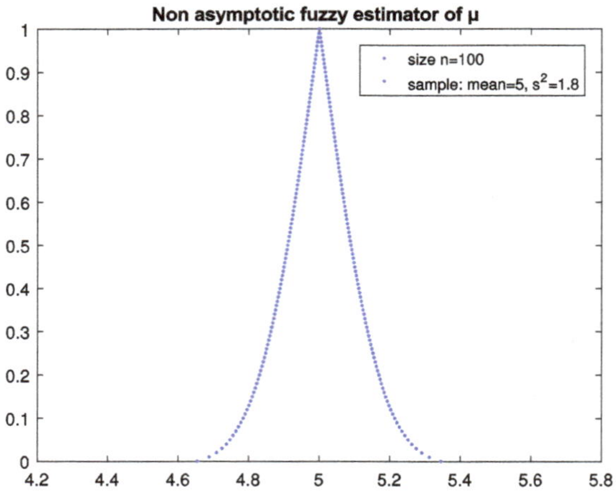

Figure 3: Nonasymptotic fuzzy estimator of μ from large sample.

4.2.3 Estimation of the mean of a normal variable with unknown variance

If the random variable X follows normal distribution with unknown variance, then the $(1-\beta)100\,\%$ confidence interval of the mean μ of X derived from a random sample of observations of X of size n and sample mean and variance $\overline{x}$ and s is [9],

$$\left[\overline{x} - t_{\frac{\beta}{2};n-1}\frac{s}{\sqrt{n}}, \overline{x} + t_{\frac{\beta}{2};n-1}\frac{t}{\sqrt{n}}\right] \tag{15}$$

where

$$t_{\frac{\beta}{2};n-1} = F^{-1}\left(1-\frac{\beta}{2}\right) \tag{16}$$

and F_{n-1}^{-1} the inverse distribution function of the t-distribution with $n-1$ degrees of freedom. So according to (7), the α-cuts of the nonasymptotic fuzzy estimator of μ are

$$\overline{\mu}[\alpha] = \left[\overline{x} - t_{h(\alpha);n-1}\frac{s}{\sqrt{n}}, \overline{x} + t_{h(\alpha);n-1}\frac{s}{\sqrt{n}}\right] \tag{17}$$

where ($h(\alpha)$ is defined in (6))

$$t_{h(\alpha);n-1} = F_{n-1}^{-1}(1-h(\alpha)) \tag{18}$$

Example 3. Implementing (17), we obtain in Figure 4 the nonasymptotic fuzzy estimator of the mean μ of a normal random variable X from a sample of $n = 25$ observations with sample mean and variance $\overline{x} = -10.7$ and $s^2 = 3.24$.

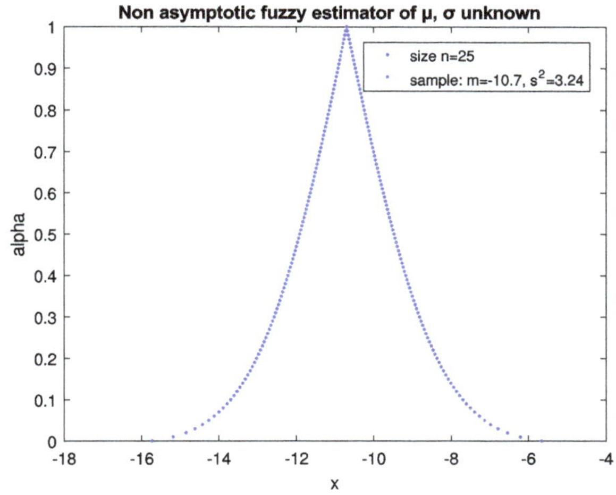

Figure 4: Nonasymptotic fuzzy estimator of μ, unknown variance.

4.2.4 Estimation of the difference $\mu_1 - \mu_2$ of the means of two normal variables with known variances

If the random variables X_1 and X_2 follow normal distribution with known variances σ_1^2 and σ_2^2, then the $(1-\beta)100\,\%$ confidence interval of the difference $\mu_1 - \mu_2$ of their means derived from two independent random samples of n_1 and n_2 observations of X_1 and X_2 with sample means $\overline{x}_1$ and $\overline{x}_2$ is [9],

$$[\overline{x}_1 - \overline{x}_2 - z_{\frac{\beta}{2}}\sigma_0, \overline{x}_1 - \overline{x}_2 + z_{\frac{\beta}{2}}\sigma_0] \tag{19}$$

where

$$\sigma_0 = \sqrt{\frac{\sigma_1^2}{n_1} + \frac{\sigma_2^2}{n_2}} \tag{20}$$

So according to (7), the α-cuts of the nonasymptotic fuzzy estimator of $\mu_1 - \mu_2$ are

$$\overline{\mu}_{12}[\alpha] = [\overline{x}_1 - \overline{x}_2 - z_{h(\alpha)}\sigma_0, \quad \overline{x}_1 - \overline{x}_2 + z_{h(\alpha)}\sigma_0], \quad 0 \leq \alpha \leq 1 \tag{21}$$

where $z_{h(\alpha)}$ is given by (12) for $h(\alpha)$ of (6).

Example 4. Implementing (21), we obtain in Figure 5 the nonasymptotic fuzzy estimator of the difference $\mu_1 - \mu_2$ of the means of two random variables X_1 and X_2 which follow normal distributions with variances $\sigma_1^2 = 10.24$ and $\sigma_2^2 = 7.84$ from two samples of $n_1 = 15$ and $n_2 = 8$ observations with sample means $\overline{x}_1 = 50.2$ and $\overline{x}_1 = 51.7$.

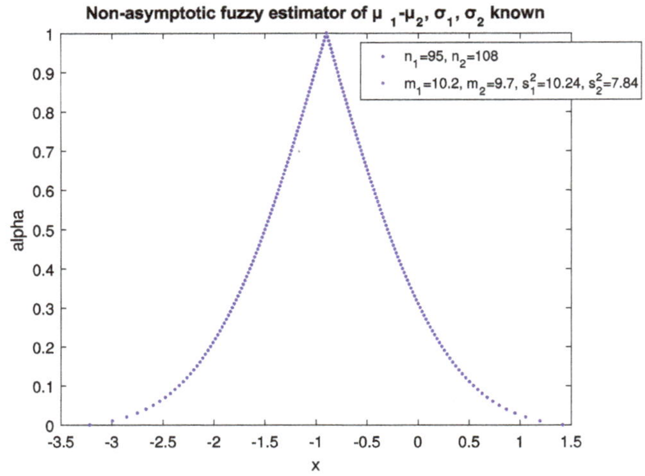

Figure 5: Nonasymptotic fuzzy estimator $\overline{\mu}_{12}$ from two samples with known variances.

4.2.5 Estimation of the difference $\mu_1 - \mu_2$ of the means of two normal variables with unknown variances from two large samples

If the random variables X_1 and X_2 follow any distribution, then the $(1-\beta)100\,\%$ confidence interval of the difference $\mu_1 - \mu_2$ of their means which is derived from two independent large random samples of n_1 and n_2 observations of X_1 and X_2 $(n_1, n_2 > 30)$ with sample means and variances $\overline{x}_1$, $\overline{x}_2$, and s_1^2, s_2^2 is [9]

$$[\overline{x}_1 - \overline{x}_2 - z_{\frac{\beta}{2}}\sigma_0, \overline{x}_1 - \overline{x}_2 + z_{\frac{\beta}{2}}\sigma_0] \tag{22}$$

where

$$\sigma_0 = \sqrt{\frac{s_1^2}{n_1} + \frac{s_2^2}{n_2}} \tag{23}$$

So according to (7), the α-cuts of the nonasymptotic fuzzy estimator $\overline{\mu}_{12}$ of $\mu_1 - \mu_2$ are

$$\overline{\mu}_{12}[\alpha] = [\overline{x}_1 - \overline{x}_2 - z_{h(\alpha)}\sigma_0, \overline{x}_1 - \overline{x}_2 + z_{h(\alpha)}\sigma_0], \quad 0 \le \alpha \le 1 \tag{24}$$

where $z_{h(\alpha)}$ is given by (12) for $h(\alpha)$ of (6).

Example 5. Implementing (24), we obtain in Figure 6 the nonasymptotic fuzzy estimator of the difference $\mu_1 - \mu_2$ of the means of two random variables X_1 and X_2, which follow any distributions, from two samples of $n_1 = 95$ and $n_2 = 108$ observations with sample means and variances $\overline{x}_1 = 10.2$, $\overline{x}_2 = 9.7$, and $s_1^2 = 3.2, s_2^2 = 2.8$.

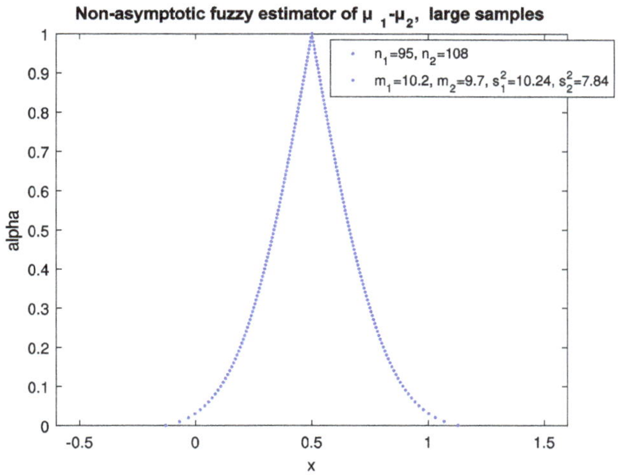

Figure 6: Nonasymptotic fuzzy estimator $\overline{\mu}_{12}$ from two large samples.

4.2.6 Estimation of the difference $\mu_1 - \mu_2$ of the means of two normal variables with unknown variances

If the random variables X_1 and X_2 follow normal distribution with unknown *equal* variances $\sigma_1^2 = \sigma_2^2$, then the $(1-\beta)100\,\%$ confidence interval of the difference $\mu_1 - \mu_2$ of the their means derived from two independent random samples of observations of X_1 and X_2 of sizes and sample means and variances $\overline{x}_1, \overline{x}_2$ and s_1^2, σ_2^2 is [9]

$$\left[\overline{x}_1 - \overline{x}_2 - t_{\frac{\beta}{2};n_1+n_2-2}\sigma_p\sqrt{\frac{1}{n_1}+\frac{1}{n_2}}, \overline{x}_1 - \overline{x}_2 + t_{\frac{\beta}{2};n_1+n_2-2}\sigma_p\sqrt{\frac{1}{n_1}+\frac{1}{n_2}}\right] \tag{25}$$

where

$$\sigma_p = \sqrt{\frac{(n_1-1)s_1^2 + (n_2-1)s_2^2}{n_1+n_2-2}} \tag{26}$$

the pooled estimator of the common variance. So according to (7), the α-cuts of the nonasymptotic fuzzy estimator $\overline{\mu}_{12}$ of $\mu_1 - \mu_2$ are

$$\begin{aligned}\overline{\mu}_{12}[\alpha] = \Bigg[&\overline{x}_1 - \overline{x}_2 - t_{h(\alpha);n_1+n_2-2}\sigma_p\sqrt{\frac{1}{n_1}+\frac{1}{n_2}},\\ &\overline{x}_1 - \overline{x}_2 + t_{h(\alpha);n_1+n_2-2}\sigma_p\sqrt{\frac{1}{n_1}+\frac{1}{n_2}}\Bigg]\end{aligned} \tag{27}$$

where $t_{h(\alpha);n_1+n_2-2}$ is given by (18) for $h(\alpha)$ of (6).

If the variances are **unequal**, then the $(1-\beta)100\,\%$ confidence interval is [9]

$$[\overline{x}_1 - \overline{x}_2 - t_{\frac{\beta}{2};r}\sigma_0, \overline{x}_1 - \overline{x}_2 + t_{\frac{\beta}{2};r}\sigma_0] \tag{28}$$

where

$$\sigma_0 = \sqrt{\frac{s_1^2}{n_1} + \frac{s_2^2}{n_2}} \tag{29}$$

and

$$r = \frac{(\frac{s_1^2}{n_1} + \frac{s_2^2}{n_2})^2}{\frac{1}{n_1-1}(\frac{s_1^2}{n_1})^2 + \frac{1}{n_2-1}(\frac{s_2^2}{n_2})^2} \tag{30}$$

So according to (7), the α-cuts of the nonasymptotic fuzzy estimator of $\mu_1 - \mu_2$ are

$$\overline{\mu}_{12}[\alpha] = \left[\overline{x}_1 - \overline{x}_2 - t_{h(\alpha);r}\sigma_0\sqrt{\frac{1}{n_1} + \frac{1}{n_2}}, \overline{x}_1 - \overline{x}_2 + t_{h(\alpha);r}\sigma_0\sqrt{\frac{1}{n_1} + \frac{1}{n_2}}\right] \tag{31}$$

Example 6. Implementing (31), we obtain in Figure 7 the nonasymptotic fuzzy estimator of the difference $\mu_1 - \mu_2$ of the means of two random variables X_1 and X_2, which follow normal distributions with equal unknown variances from two samples of $n_1 = 25$ and $n_2 = 18$ observations with sample means and variances $\overline{x}_1 = 10.2$, $\overline{x}_2 = 9.7$, and $s_1^2 = 3.2, s_2^2 = 2.8$.

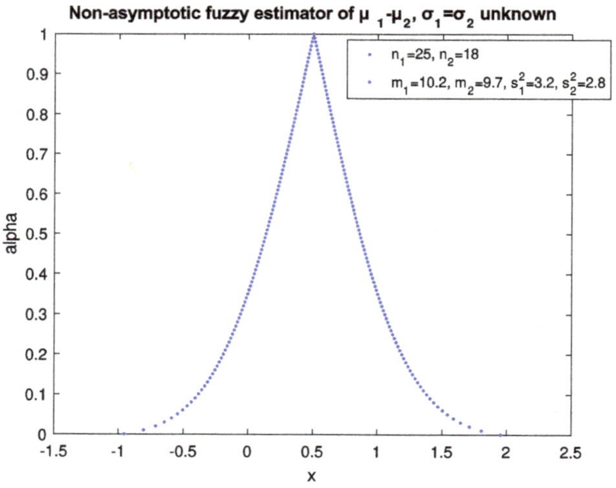

Figure 7: Nonasymptotic fuzzy estimator $\overline{\mu}_{12}$ from two small samples with unknown variances.

4.2.7 Estimation of the probability of success p of a binomial distribution

It is known that if n is sufficiently large ($n > 30$), then using the normal approximation to the binomial, the $(1-\beta)100\,\%$ confidence interval of the probability of success p of a binomial distribution from a random sample of n observations in which happened x successes is [9],

$$\left[\widehat{p} - z_{\frac{\beta}{2}}\sqrt{\frac{\widehat{p}\widehat{q}}{n}}, \overline{x} + z_{\frac{\beta}{2}}\sqrt{\frac{\widehat{p}\widehat{q}}{n}}\right] \tag{32}$$

where $\widehat{p} = \frac{x}{n}$ and $\widehat{q} = 1 - \widehat{p}$. So according to (7), the α-cuts of the nonasymptotic fuzzy estimator of p are

$$\overline{p}[\alpha] = \left[\widehat{p} - z_{h(\alpha)}\sqrt{\frac{\widehat{p}\widehat{q}}{n}}, \widehat{p} + z_{h(\alpha)}\sqrt{\frac{\widehat{p}\widehat{q}}{n}}\right], \quad 0 \leq \alpha \leq 1 \tag{33}$$

where $z_{h(\alpha)}$ is given by (12) for $h(\alpha)$ of (6).

Example 7. Implementing (33), we obtain in Figure 8 the nonasymptotic fuzzy estimator of the probability of success p of a binomial distribution from a large sample of $n = 200$ observations in which happened $x = 124$ successes.

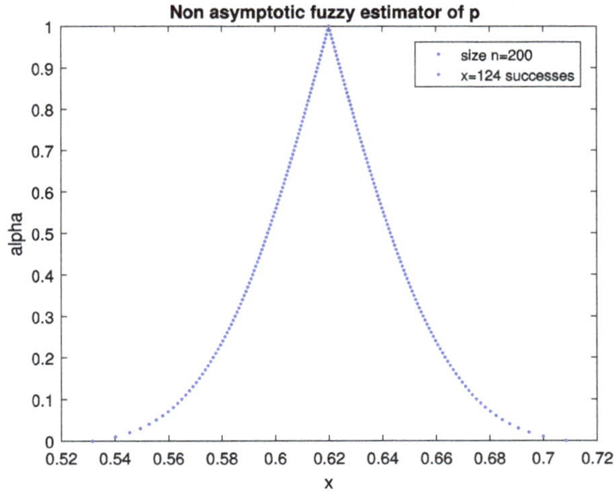

Figure 8: Nonasymptotic fuzzy estimator $\overline{p}$ of the probability of success of a binomial distribution.

4.2.8 Estimation of the difference $p_1 - p_2$ of the probabilities of success of two binomial distributions

It is known that if n_1 and n_2 are sufficiently large ($n_1, n_2 > 30$), then using the normal approximation to the binomial, the $(1-\beta)100\,\%$ confidence interval of the probability of the difference $p_1 - p_2$ of the probabilities of success of two binomial distributions from two random sample of n_1 and n_2 observations in which happened x_1 and x_2 successes is [9],

$$\left[\widehat{p}_1 - \widehat{p}_2 - z_{\frac{\beta}{2}}\sqrt{\frac{\widehat{p}_1\widehat{q}_1}{n_1} + \frac{\widehat{p}_2\widehat{q}_2}{n_2}}, \widehat{p}_1 - \widehat{p}_2 + z_{\frac{\beta}{2}}\sqrt{\frac{\widehat{p}_1\widehat{q}_1}{n_1} + \frac{\widehat{p}_2\widehat{q}_2}{n_2}}\right] \tag{34}$$

where

$$\widehat{p}_1 = \frac{x_1}{n_1}, \quad \widehat{p}_2 = \frac{x_2}{n_2}, \quad \widehat{q}_1 = 1 - \widehat{p}_1, \quad \widehat{q}_2 = 1 - \widehat{p}_2$$

So according to (7), the α-cuts of the nonasymptotic fuzzy estimator of p are

$$\overline{p}_{12}[\alpha] = \left[\widehat{p}_1 - \widehat{p}_2 - z_{h(\alpha)}\sqrt{\frac{\widehat{p}_1\widehat{q}_1}{n_1} + \frac{\widehat{p}_2\widehat{q}_2}{n_2}}, \widehat{p}_1 - \widehat{p}_2 + z_{h(\alpha)}\sqrt{\frac{\widehat{p}_1\widehat{q}_1}{n_1} + \frac{\widehat{p}_2\widehat{q}_2}{n_2}}\right] \tag{35}$$

where $z_{h(\alpha)}$ is given by (12) for $h(\alpha)$ of (6).

Example 8. Implementing (35), we obtain in Figure 9 the nonasymptotic fuzzy estimator of the difference $p_1 - p_2$ of the probabilities of success p_1, p_2 of two binomial distributions from two random sample of $n_1 = 250$ and $n_2 = 400$ observations in which happened $x_1 = 40$ and $x_2 = 55$ successes.

4.2.9 Estimation of the variance of a normal variable

The $(1-\beta)100\,\%$ confidence interval of the variance σ^2 of a normal random variable X derived from a random sample of n observations of X with sample variance s^2 is [9]

$$\left[\frac{(n-1)s^2}{\chi^2_{R,\frac{\beta}{2};n-1}}, \quad \frac{(n-1)s^2}{\chi^2_{L,\frac{\beta}{2};n-1}}\right]$$

where (F_{n-1}^{-1} the inverse distribution function of the χ^2_{n-1}-distribution)

$$\chi^2_{L,h(\alpha);n-1} = F_{n-1}^{-1}(h(\alpha)) \text{ and } \chi^2_{R,h(\alpha);n-1} = F_{n-1}^{-1}(1-h(\alpha)), \quad 0 \le \alpha \le 1 \tag{36}$$

So according to (7), the α-cuts of the nonasymptotic fuzzy estimator $\overline{\sigma^2}$ of the variance σ^2 of X are

$$\overline{\sigma^2}[\alpha] = \left[\frac{(n-1)s^2}{\chi^2_{R,h(\alpha);n-1}}, \quad \frac{(n-1)s^2}{\chi^2_{L,h(\alpha);n-1}}\right] \tag{37}$$

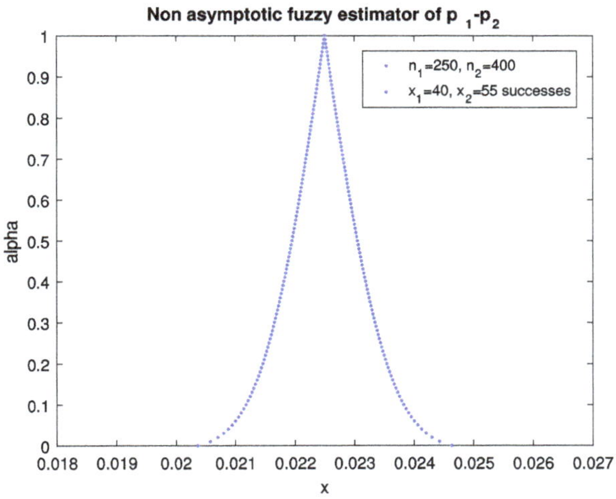

Figure 9: Nonasymptotic fuzzy estimator $\overline{p}_{12}$ of the difference $p_1 - p_2$ of the probabilities of success of two binomial distributions.

where $h(\alpha)$ is defined in (6). Therefore, the α-cuts of the nonasymptotic fuzzy estimator of the standard deviation σ are

$$\overline{\sigma}[\alpha] = \left[s\sqrt{\frac{(n-1)}{\chi^2_{R;h(\alpha);n-1}}},\quad s\sqrt{\frac{(n-1)}{\chi^2_{L;1-h(\alpha);n-1}}} \right], \quad 0 \le \alpha \le 1 \tag{38}$$

It can be shown that the fuzzy estimator $\overline{\sigma^2}$ of (37) is biased, because its core is not at s^2. A fuzzy estimator of a parameter is defined as biased when its core is not at the crisp point estimator of the parameter.

In [5], an unbiased fuzzy estimator $\overline{\sigma^2}_u$ of the variance σ^2 of a normal random variable from a sample of size n and variance s^2 is obtained by putting one above the other the confidence intervals

$$\left[\frac{(n-1)s^2}{L(\lambda)},\quad \frac{(n-1)s^2}{R(\lambda)} \right], \quad 0 \le \lambda \le 1$$

where

$$L(\lambda) = (1-\lambda)\chi^2_{R,0.005;n-1} + (n-1)\lambda$$
$$R(\lambda) = (1-\lambda)\chi^2_{L,0.005;n-1} + (n-1)\lambda$$

Example 9. Implementing (37), we obtain in Figure 10 the nonasymptotic fuzzy estimator of the variance σ^2 of a normal random variable X derived from a sample of $n = 25$ observations with sample variance $s^2 = 1.2$.

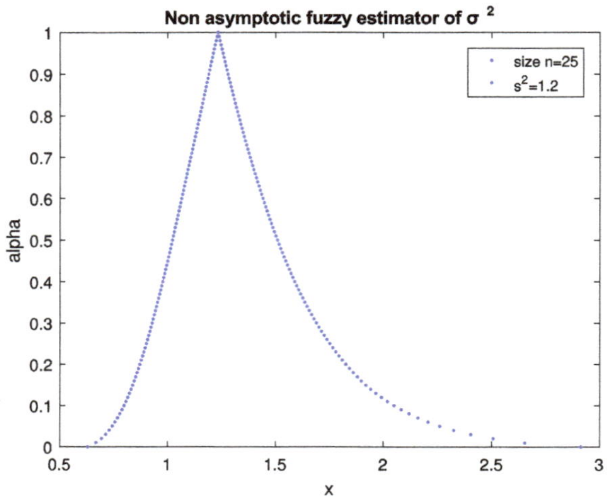

Figure 10: Nonasymptotic fuzzy estimator $\overline{\sigma^2}$ of the variance of a normal random variable.

4.2.10 Estimation of σ_1^2/σ_2^2 for the variances of two normal distributions

The $(1-\beta)100\,\%$ confidence interval of the ratio σ_1^2/σ_2^2 of the variances of two normal random variables X_1 and X_2 derived from two independent random samples of n_1 and n_2 observations of X_1 and X_2 with sample variances s_1^2 and s_2^2 is [9],

$$\left[F_{1-\gamma/2;n_1-1,n_2-1}\frac{s_1^2}{s_2^2}, F_{\gamma/2;n_1-1,n_2-1}\frac{s_1^2}{s_2^2}\right] \tag{39}$$

where

$$F_{\gamma/2;n_1-1,n_2-1} = F^{-1}\left(1-\frac{\gamma}{2}\right) \quad \text{and} \quad F_{1-\gamma/2;n_1-1,n_2-1} = F^{-1}\left(\frac{\gamma}{2}\right) \tag{40}$$

So according to (7), the α-cuts of the nonasymptotic fuzzy estimator of σ_1^2/σ_2^2 are

$$\overline{\sigma^2}_{12}[\alpha] = \left[F_{1-h(\alpha);n_1-1,n_2-1}\frac{s_1^2}{s_2^2}, F_{h(\alpha);n_1-1,n_2-1}\frac{s_1^2}{s_2^2}\right], \quad 0 \le \alpha \le 1 \tag{41}$$

where $h(\alpha)$ is given by (6).

Example 10. Implementing (41), we obtain in Figure 11 the nonasymptotic fuzzy estimator of σ_1^2/σ_2^2 for the variances of two normal random variables X_1 and X_2 from two samples of $n_1 = 11$ and $n_2 = 13$ observations with sample variances $\overline{x}_1 = 0.24$ and $\overline{x}_1 = 0.8$.

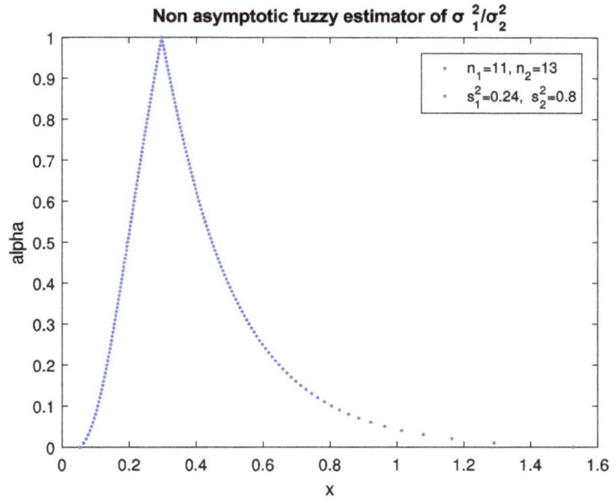

Figure 11: Nonasymptotic fuzzy estimator $\overline{\sigma^2}_{12}$.

4.3 Ordering fuzzy numbers

The fuzzy hypotheses tests are based on ordering fuzzy numbers, for which we will use one of the several procedures used (as in [1, 3, 6, 7, 11]), according to which the degree of the inequality $A \le B$ that counts the degree at which the fuzzy number A is less or equal to the fuzzy number B is defined as [5],

$$v(A \le B) = \max\{\min(A(x), B(y)), x \le y\} \tag{42}$$

According to (42):

If $v(A \le B) = 1$, then we define the degree of confidence (truth-value) d of $A < B$ as

$$v(A < B) = d \Leftrightarrow v(A \le B) = 1 \quad \text{and} \quad v(B \le A) = 1 - d \tag{43}$$

If $v(B \le A) = 1$, then the truth-value d of $B < A$ is defined as

$$v(B < A) = d \Leftrightarrow v(B \le A) = 1 \quad \text{and} \quad v(A \le B) = 1 - d \tag{44}$$

The truth-value η of $A \approx B$ is defined as

$$\begin{aligned} v(A \approx B) = \eta \Leftrightarrow v(A \le B) = 1 \quad &\text{and} \quad v(B \le A) = \eta \\ \text{or} \quad v(B \le A) = 1 \text{ and } v(A \le B) = \eta & \end{aligned} \tag{45}$$

In the case of the triangular shaped fuzzy numbers of Figure 12, which appears often in fuzzy hypotheses tests, we can see that:

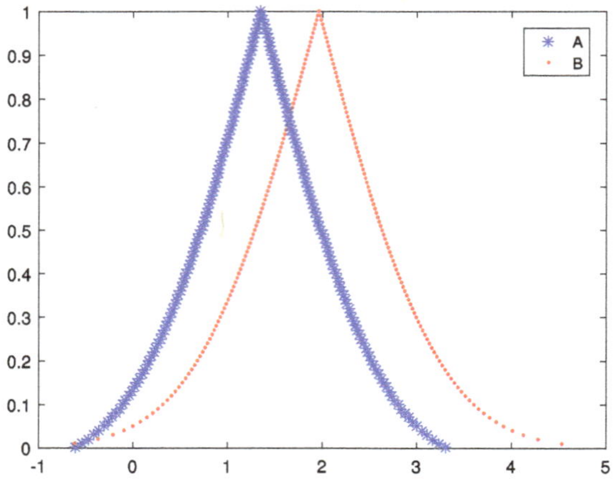

Figure 12: Ordering triangular shaped fuzzy numbers.

> $v(A \le B) = 1$ according to (42), since the core of A lies to the left of the core of B;
> $v(B \le A) = y_0$ according to (42), where y_0 the truth level of the point of intersection of the right part of A and the left part of B.

Thus according to (43) and (45),

$$v(A < B) = 1 - y_0 \quad \text{and} \quad v(A \approx B) = y_0$$

4.4 Hypotheses testing using fuzzy test statistics and fuzzy critical values or fuzzy *p*-value

In a fuzzy hypothesis test of a null hypothesis $H_0 : \theta = \theta_0$ for a parameter θ of the distribution of a random variable x, we use a fuzzy test statistic $\overline{U}$, which is produced by substituting the crisp estimator of θ in the respective crisp statistic U with a fuzzy estimator $\overline{\theta}$. As in the crisp case, there are two methods for such a test:

(1) We find the fuzzy test statistic $\overline{U}$ and the fuzzy critical values $\overline{\mathrm{CV}}_i$ and reject or accept H_0 comparing the fuzzy numbers $\overline{U}$ and $\overline{\mathrm{CV}}_i$, as described in Section 4.4.1;

(2) We find the fuzzy p-value $\overline{P}$, fuzzify the significance level γ to $\overline{S}$ and reject or accept H_0 comparing the fuzzy numbers $\overline{P}$ and $\overline{S}$, as described in Section 4.4.2.

In Sections 4.5–4.10, we apply these methods in several cases of fuzzy hypotheses tests.

4.4.1 Using fuzzy critical values

In a fuzzy hypothesis test of $H_0 : \theta = \theta_0$ at significance level γ, the critical values are the fuzzy numbers $\overline{\mathrm{CV}}_1$ and $\overline{\mathrm{CV}}_2$ the left and right a-cuts of which $\mathrm{CV}_{1l}[a], \mathrm{CV}_{1r}[a], \mathrm{CV}_{2l}[a]$, $\mathrm{CV}_{2r}[a]$ are defined in [5] as follows:

(a) For **a one-sided test from the left** (alternative hypothesis $H_1 : \theta < \theta_0$) according to (1):

$$P(\overline{U}_l[a] \le \overline{\mathrm{CV}}_{1l}[a]) = \gamma \tag{46}$$

$$P(\overline{U}_r[a] \le \overline{\mathrm{CV}}_{1r}[a]) = \gamma \tag{47}$$

(b) For **a one-sided test from the right** (alternative hypothesis $H_1 : \theta > \theta_0$) according to (2):

$$P(\overline{U}_l[a] \ge \overline{\mathrm{CV}}_{2l}[a]) = \gamma \tag{48}$$

$$P(\overline{U}_r[a] \ge \overline{\mathrm{CV}}_{2r}[a]) = \gamma \tag{49}$$

(c) For **a two-sided test** (alternative hypothesis $H_1 : \theta \ne \theta_0$) according to (3):

$$P(\overline{U}_l[a] \le \overline{\mathrm{CV}}_{1l}[a]) = \frac{\gamma}{2} \tag{50}$$

$$P(\overline{U}_r[a] \le \overline{\mathrm{CV}}_{1r}[a]) = \frac{\gamma}{2} \tag{51}$$

$$P(\overline{U}_l[a] \ge \overline{\mathrm{CV}}_{2l}[a]) = \frac{\gamma}{2} \tag{52}$$

$$P(\overline{U}_r[a] \ge \overline{\mathrm{CV}}_{2r}[a]) = \frac{\gamma}{2} \tag{53}$$

where $\overline{U}_l[a], \overline{U}_r[a]$ the left and right a-cuts of the fuzzy test statistic $\overline{U}$.

$\overline{\mathrm{CV}}_{1l}[a], \overline{\mathrm{CV}}_{1r}[a]$, and $\overline{\mathrm{CV}}_{2l}[a], \overline{\mathrm{CV}}_{2r}[a]$ are found in each case using the distribution followed by the test statistic U, as shown in Sections 4.5–4.10.

The decision for rejecting or not H_0 from a given random sample at significance level γ, depends on the ordering of the fuzzy numbers $\overline{U}_0$ (the fuzzy test statistic $\overline{U}$ for the given sample) and $\overline{\mathrm{CV}}_i$, as described in Section 4.3. So:

(a) For **a one-sided test from the left** (alternative hypothesis $H_1 : \theta < \theta_0$):
If $v(\overline{U}_0 < \overline{\mathrm{CV}}_1) = d$, then H_0 is rejected with rejection degree d.
If $v(\overline{U}_0 > \overline{\mathrm{CV}}_1) = d$, then H_0 is accepted with acceptance degree d.
If $v(\overline{U}_0 \approx \overline{\mathrm{CV}}_1) = d_1$ (close to 1), then we cannot make a decision on rejecting or not H_0 with any degree of confidence greater or equal to $1 - d_1$.

(b) For **a one-sided test from the right** (alternative hypothesis $H_1 : \theta > \theta_0$):
If $v(\overline{U}_0 > \overline{\mathrm{CV}}_2) = d$, then H_0 is rejected with degree of rejection d.
If $v(\overline{U}_0 < \overline{\mathrm{CV}}_2) = d$, then H_0 is accepted with degree of acceptance d.
If $v(\overline{U}_0 \approx \overline{\mathrm{CV}}_2) = d_1$ (close to 1), then we cannot make a decision on rejecting or not H_0 with any degree of confidence greater or equal to $1 - d_1$.

(c) For **a two-sided test** (alternative hypothesis $H_1 : \theta \neq \theta_0$):
If $\max(v(\overline{U}_0 < \overline{\mathrm{CV}}_1), v(\overline{U}_0 > \overline{\mathrm{CV}}_2)) = d$, then H_0 is rejected with degree of rejection d.
If $\min(v(\overline{U}_0 > \overline{\mathrm{CV}}_1), v(\overline{U}_0 < \overline{\mathrm{CV}}_2)) = d$, then H_0 is accepted with degree of acceptance d.
If $\max(v(\overline{U}_0 \approx \overline{\mathrm{CV}}_2), v(\overline{U}_0 \approx \overline{\mathrm{CV}}_1)) = d_1$ (close to 1), then we cannot make a decision on rejecting or not H_0 with any degree of confidence greater or equal to $1 - d_1$.

4.4.2 Using fuzzy *p*-value

Using Zadeh's extension principle [20], a fuzzy p-value is defined as a fuzzy number, the α-cuts of which, as described in [8], are:

(a) For **a one-sided test from the left** (alternative hypothesis $H_1 : \theta < \theta_0$):

$$\overline{P}[\alpha] = [\Pr(U \leq \overline{U}_l[\alpha]), \Pr(U \leq \overline{U}_r[\alpha])] \tag{54}$$

where U the test statistic and $\overline{U}_l[\alpha]$ and $\overline{U}_r[\alpha]$ the α-cuts of the fuzzy test statistic $\overline{U}$ for the given sample.

(b) For **a one-sided test from the right** (alternative hypothesis $H_1 : \theta > \theta_0$):

$$\overline{P}[\alpha] = [\Pr(U \geq \overline{U}_r[\alpha]), \Pr(U \geq \overline{U}_l[\alpha])] \tag{55}$$

(c) For **a two-sided test** (alternative hypothesis $H_1 : \theta \neq \theta_0$):

$$\overline{P}[\alpha] = \begin{cases} [2\Pr(U \leq \overline{U}_l[\alpha]), \min\{1, 2(\Pr(U \leq \overline{U}_r[\alpha]))\}], & u_0 < m_U \\ [2\Pr(U \geq \overline{U}_r[\alpha]), \min\{1, 2\Pr(U \geq \overline{U}_l[\alpha])\}], & u_0 \geq m_U \end{cases} \tag{56}$$

where m_U the median of the distribution of U under the null hypothesis H_0 and u_0 the core of the fuzzy test statistic $\overline{U}$ (where $\overline{U}(u_0) = 1$).

Since the *p*-value is fuzzy, the significance level has to be a fuzzy set $\overline{S}$. A simple way to fuzzify a significance level γ is to use a triangular fuzzy numbers of one of the forms [10]

$$\overline{S}(a) = T(0, \gamma, 2\gamma) \quad \text{or} \quad \overline{S}(a) = T(\gamma - c, \gamma, 2\gamma + c), \quad c \text{ constant} \tag{57}$$

which expresses the notion "the significance level is approximately γ." So using fuzzy ordering between $\overline{P}$ and $\overline{S}$ (see Section 4.3):

If $v(\overline{P} < \overline{S}) = d$, then H_0 is rejected with degree of rejection d.
If $v(\overline{P} > \overline{S}) = d$, then H_0 is accepted with degree of acceptance d.
If $v(\overline{P} \approx \overline{S}) = d_1$ (close to 1), then we cannot make a decision on rejecting or not H_0 with any degree of confidence greater or equal to $1 - d_1$.

4.5 Tests on the mean of a normal distribution with known variance

We test at significance level γ the null hypothesis

$$H_0 : \mu = \mu_0$$

for the mean μ of a random variable X, which follows normal distribution with known variance σ, using a random sample of observations of X of size n.

4.5.1 Using fuzzy critical values

In the crisp case, we test H_0 using the statistic [9],

$$Z = \frac{\overline{X} - \mu_0}{\sigma/\sqrt{n}} \tag{58}$$

where $\overline{X}$ is the statistic of the sample mean. It is known that under the null hypothesis ($\mu = \mu_0$) Z follows the standard normal distribution $N(0,1)$, so H_0 is rejected from a given sample:

(a) for the *one-sided test from the right*, if

$$z_0 \geq z_\gamma, \quad \text{where } z_\gamma = \Phi^{-1}(1-\gamma),$$

(b) for the *one-sided test from the left*, if $z_0 \leq -z_\gamma = \Phi^{-1}(\gamma)$,

(c) for the *two-sided* test, if $z_0 \leq -z_{\gamma/2}$ or $z_0 \geq z_{\gamma/2}$, where

$$z_0 = \frac{\overline{x} - \mu_0}{\sigma/\sqrt{n}} \tag{59}$$

the value of the statistic (58) for the given sample,

$$z_c = z_{\gamma/2} = \Phi^{-1}\left(1 - \frac{\gamma}{2}\right) \tag{60}$$

the critical value of the test and Φ^{-1} the inverse distribution function of the standard normal distribution. While, H_0 is not rejected if $-z_c < z_0 < z_c$.

In the fuzzy case, the test of H_0 is based on the fuzzy statistic [5],

$$\overline{Z} = \frac{\overline{\mu} - \mu_0}{\sigma/\sqrt{n}} \tag{61}$$

which is generated by substituting $\overline{X}$ in (58) with a fuzzy estimator $\overline{\mu}$ of the mean value for the given sample, the α-cuts of which are found in Section 4.2.1 to be

$$\overline{\mu}[\alpha] = \left[\overline{X} - z_{h(\alpha)}\frac{\sigma}{\sqrt{n}}, \overline{X} + z_{h(\alpha)}\frac{\sigma}{\sqrt{n}}\right], \quad \alpha \in [0,1] \tag{62}$$

where $h(\alpha)$ and $z_{h(\alpha)}$ are given by (6) and (12).

From (58), (61), (62), and interval arithmetic follows that the α-cuts of the fuzzy statistic $\overline{Z}$ are

$$\overline{Z}[\alpha] = [Z - z_{h(\alpha)}, Z + z_{h(\alpha)}], \quad \alpha \in [0,1]$$

Hence, for the given sample we get the fuzzy number

$$\overline{Z}_0[\alpha] = [z_0 - z_{h(\alpha)}, z_0 + z_{h(\alpha)}], \quad \alpha \in [0,1] \tag{63}$$

where z_0 the value of the statistic Z for this sample which is given by (59).

This generalizes to a function the fuzzy statistic $\overline{Z}_0$ of [5], the α-cuts of which are

$$\overline{Z}_0[\alpha] = \begin{cases} [z_0 - z_{\alpha/2}, z_0 + z_{\alpha/2}], & 0.01 \le \alpha \le 1 \\ [z_0 - z_{0.005}, z_0 + z_{0.005}], & 0 \le \alpha < 0.01 \end{cases}$$

The critical values of this test are the fuzzy numbers $\overline{\mathrm{CV}}_1$ and $\overline{\mathrm{CV}}_2$, the a-cuts of which are found as described in Section 4.4.1 ($\overline{Z}_l[a], \overline{Z}_r[a]$, $\overline{\mathrm{CV}}_{1l}[a], \overline{\mathrm{CV}}_{1r}[a]$, and $\overline{\mathrm{CV}}_{2l}[a], \overline{\mathrm{CV}}_{2r}[a]$ the left and right α-cuts of $\overline{Z}$, $\overline{\mathrm{CV}}_1$, and $\overline{\mathrm{CV}}_2$):

1. For the *one-sided test from the left* (alternative hypothesis $H_1 : \mu < \mu_0$) according to (46),

$$P(\overline{Z}_l[a] \le \overline{\mathrm{CV}}_{1l}[a]) = \gamma \Leftrightarrow P(Z - z_{h(\alpha)} \le \overline{\mathrm{CV}}_{1l}[a]) = \gamma$$
$$\Leftrightarrow P(Z \le \overline{\mathrm{CV}}_{1l}[a] + z_{h(\alpha)}) = \gamma$$

So, since the test statistic Z follows the standard normal distribution,

$$\overline{\mathrm{CV}}_{1l}[a] + z_{h(\alpha)} = -z_\gamma \Leftrightarrow \overline{\mathrm{CV}}_{1l}[a] = -z_\gamma - z_{h(\alpha)}$$

Similarly, according to (47),

$$P(\overline{Z}_r[a] \le \overline{\mathrm{CV}}_{1r}[a]) = \gamma \Leftrightarrow P(Z + z_{h(\alpha)} \le \overline{\mathrm{CV}}_{1r}[a]) = \gamma$$
$$\Leftrightarrow P(Z \le \overline{\mathrm{CV}}_{1r}[a] - z_{h(\alpha)}) = \gamma$$

So, since the test statistic Z follows the standard normal distribution,

$$\overline{\mathrm{CV}}_{1r}[a] - z_{h(\alpha)} = -z_\gamma \Leftrightarrow \overline{\mathrm{CV}}_{1r}[a] = -z_\gamma + z_{h(\alpha)}$$

Therefore, the α-cuts of $\overline{\mathrm{CV}}_1$ for the one-sided test from the left are

$$\overline{\mathrm{CV}}_1[\alpha] = [-z_\gamma - z_{h(\alpha)}, -z_\gamma + z_{h(\alpha)}] \tag{64}$$

2. For the **one-sided test from the right** (alternative hypothesis $H_1 : \mu > \mu_0$) according to (48),

$$P(\overline{Z}_l[a] \geq \overline{\mathrm{CV}}_{2l}[a]) = \gamma \Leftrightarrow P(Z - z_{h(\alpha)} \geq \overline{\mathrm{CV}}_{2l}[a]) = \gamma$$
$$\Leftrightarrow P(Z \geq z_{h(\alpha)} + \overline{\mathrm{CV}}_{2l}[a]) = \gamma$$

So, since the test statistic Z follows the standard normal distribution,

$$z_{h(\alpha)} + \overline{\mathrm{CV}}_{2l}[a] = z_\gamma \Leftrightarrow \overline{\mathrm{CV}}_{2l}[a] = z_\gamma - z_{h(\alpha)}$$

Similarly, according to (49),

$$P(\overline{Z}_r[a] \geq \overline{\mathrm{CV}}_{2r}[a]) = \gamma \Leftrightarrow P(Z + z_{h(\alpha)} \geq \overline{\mathrm{CV}}_{2r}[a]) = \gamma$$
$$\Leftrightarrow P(Z \geq \overline{\mathrm{CV}}_{2r}[a] - z_{h(\alpha)}) = \gamma$$

So, since the test statistic Z follows the standard normal distribution,

$$\overline{\mathrm{CV}}_{2r}[a] - z_{h(\alpha)} = z_\gamma \Leftrightarrow \overline{\mathrm{CV}}_{2r}[a] = z_\gamma + z_{h(\alpha)}$$

Therefore, the α-cuts of $\overline{\mathrm{CV}}_2$ for the one-sided test from the right are

$$\overline{\mathrm{CV}}_2[\alpha] = [z_\gamma - z_{h(\alpha)}, z_\gamma + z_{h(\alpha)}], \quad \alpha \in [0,1] \tag{65}$$

3. For the **two-sided test** (alternative hypothesis $H_1 : \mu \neq \mu_0$) according to (52),

$$P(\overline{Z}_l[a] \geq \overline{\mathrm{CV}}_{2l}[a]) = \frac{\gamma}{2} \Leftrightarrow P(Z - z_{h(\alpha)} \geq \overline{\mathrm{CV}}_{2l}[a]) = \frac{\gamma}{2}$$
$$\Leftrightarrow P(Z \geq z_{h(\alpha)} + \overline{\mathrm{CV}}_{2l}[a]) = \frac{\gamma}{2}$$

So, since the test statistic Z follows the standard normal distribution,

$$z_{h(\alpha)} + \overline{\mathrm{CV}}_{2l}[a] = z_{\gamma/2} \Leftrightarrow \overline{\mathrm{CV}}_{2l}[a] = z_{\gamma/2} - z_{h(\alpha)}$$

Similarly, according to (53),

$$P(\overline{Z}_r[a] \geq \overline{\mathrm{CV}}_{2r}[a]) = \frac{\gamma}{2} \Leftrightarrow P(Z + z_{h(\alpha)} \geq \overline{\mathrm{CV}}_{2r}[a]) = \frac{\gamma}{2}$$
$$\Leftrightarrow P(Z \geq \overline{\mathrm{CV}}_{2r}[a] - z_{h(\alpha)}) = \frac{\gamma}{2}$$

So, since the test statistic Z follows the standard normal distribution,

$$\overline{\mathrm{CV}}_{2r}[a] - z_{h(\alpha)} = z_{\gamma/2} \Leftrightarrow \overline{\mathrm{CV}}_{2r}[a] = z_{\gamma/2} + z_{h(\alpha)}$$

In the same way according to (50),

$$P(\overline{Z}_l[a] \leq \overline{\mathrm{CV}}_{1l}[a]) = \frac{\gamma}{2} \Leftrightarrow P(Z - z_{h(\alpha)} \leq \overline{\mathrm{CV}}_{1l}[a]) = \frac{\gamma}{2}$$
$$\Leftrightarrow P(Z \leq \overline{\mathrm{CV}}_{1l}[a] + z_{h(\alpha)}) = \frac{\gamma}{2}$$

So, since the test statistic Z follows the standard normal distribution,

$$\overline{\mathrm{CV}}_{1l}[a] + z_{h(\alpha)} = -z_{\gamma/2} \Leftrightarrow \overline{\mathrm{CV}}_{1l}[a] = -z_{\gamma/2} - z_{h(\alpha)}$$

Similarly, according to (51),

$$P(\overline{Z}_r[a] \leq \overline{\mathrm{CV}}_{1r}[a]) = \frac{\gamma}{2} \Leftrightarrow P(Z + z_{h(\alpha)} \leq \overline{\mathrm{CV}}_{1r}[a]) = \frac{\gamma}{2}$$
$$\Leftrightarrow P(Z \leq \overline{\mathrm{CV}}_{1r}[a] - z_{h(\alpha)}) = \frac{\gamma}{2}$$

So, since the test statistic Z follows the standard normal distribution,

$$\overline{\mathrm{CV}}_{1r}[a] - z_{h(\alpha)} = -z_{\gamma/2} \Leftrightarrow \overline{\mathrm{CV}}_{1r}[a] = -z_{\gamma/2} + z_{h(\alpha)}$$

Therefore, the α-cuts of $\overline{\mathrm{CV}}_1$ and $\overline{\mathrm{CV}}_2$ for the two-sided test are

$$\begin{aligned} \overline{\mathrm{CV}}_1[\alpha] &= [-z_{\gamma/2} - z_{h(\alpha)}, -z_{\gamma/2} + z_{h(\alpha)}] \\ \overline{\mathrm{CV}}_2[\alpha] &= [z_{\gamma/2} - z_{h(\alpha)}, z_{\gamma/2} + z_{h(\alpha)}], \quad \alpha \in [0,1] \end{aligned} \tag{66}$$

Our decision for rejecting or not H_0 from a given random sample depends on the ordering of the fuzzy numbers $\overline{Z}_0$ (the fuzzy test statistic $\overline{Z}$ for the given sample) and $\overline{\mathrm{CV}}_i$, as described in Section 4.4.1.

Example 11.

(a) We test at significance level $\gamma = 0.05$ the null hypothesis $H_0 : \mu = 1$ for the mean μ of a random variable X, which follows normal distribution with variance $\sigma^2 = 4$ from a sample of $n = 25$ observations with sample mean $\overline{x} = 1.24$ with alternative:
  (i) $H_1 : \mu \neq 1$ (two-sided test)
  (ii) $H_1 : \mu > 1$ (one-sided test from the right)
  using fuzzy critical values.

(b) We apply the above tests changing only the sample mean to $\overline{x} = 1.71$.

(a) The value of the test statistic (58) is evaluated by (59),

$$z_0 = \frac{1.24 - 1}{2/\sqrt{25}} = 0.6$$

Since $z_0 = 0.6 < z_{0.05/2} = 1.96$, H_0 is not rejected by this crisp test.

(i) We apply the fuzzy two-sided test of H_0 implementing (63) and (66). So we obtain the results of Figure 13, where the core of $\overline{Z}$ is between the cores of $\overline{\mathrm{CV}}_1$ and $\overline{\mathrm{CV}}_2$ and the point of intersection of $\overline{Z}$ and $\overline{\mathrm{CV}}_1$ has $y_1 = 0.2$ and of $\overline{Z}$ and $\overline{\mathrm{CV}}_2$, $y_2 = 0.51$. Hence,

$$\min((\overline{Z} > \overline{\mathrm{CV}}_1), v(\overline{Z} < \overline{\mathrm{CV}}_2)) = \min(1 - y_1, 1 - y_2) = 0.49,$$

so H_0 is accepted by this test with degree of acceptance $d = 0.49$.

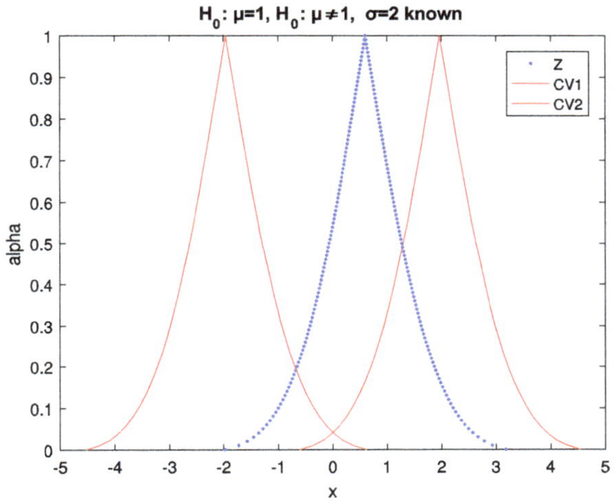

Figure 13: Fuzzy statistic $\overline{Z}$ and critical values $\overline{\mathrm{CV}}_i$ for the two-sided test of Example 11a.

(ii) We apply the one-sided from the right fuzzy test of H_0 implementing (63) and (65). So, we obtain the results of Figure 14, where the core of $\overline{Z}$ is at the right of the core of $\overline{\mathrm{CV}}_2$ and their point of intersection has $y_2 = 0.62$. Hence,

$$v(\overline{Z} < \overline{\mathrm{CV}}_2) = 1 - y_2 = 0.38,$$

so H_0 is accepted by this test with degree of acceptance $d = 0.38$.

(b) The value of the test statistic (58) in this case is

$$z_0 = \frac{1.71 - 1}{2/\sqrt{25}} = 1.775$$

Since $z_0 = 1.775 < z_{0.05/2} = 1.96$, H_0 is not rejected by this crisp test.

(i) We apply the fuzzy two-sided test of H_0 implementing (63) and (66). So we obtain the results of Figure 15, where the point of intersection of $\overline{Z}$ and $\overline{\mathrm{CV}}_2$ has $y_2 = 0.95$. Hence, according to (45),

$$v(\overline{Z} \approx \overline{\mathrm{CV}}_2) = 0.95$$

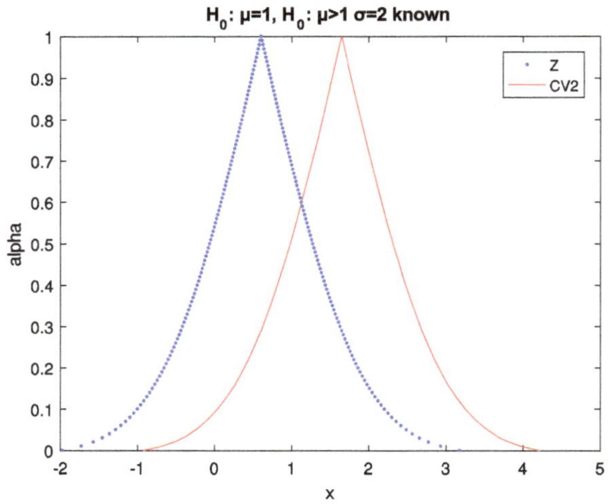

Figure 14: Fuzzy statistic $\overline{Z}$ and critical values $\overline{\text{CV}}_i$ for the one-sided test of Example 11a.

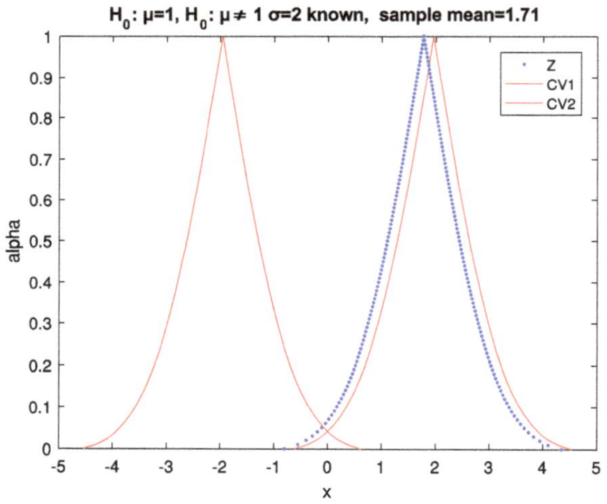

Figure 15: Fuzzy statistic $\overline{Z}$ and critical values $\overline{\text{CV}}_i$ for the two-sided test of Example 11b.

so (see Section 4.4.1) we cannot make a decision on the rejection or not of H_0 with any degree of acceptance $d > 0.05$.

(ii) We apply the fuzzy one-sided from the right test of H_0 implementing (63) and (65). So, we obtain similar results (the point of intersection of $\overline{Z}$ and $\overline{\text{CV}}_2$ has $y_0 = 0.94$),

$$v(\overline{Z} \approx \overline{\text{CV}}_2) = 0.94$$

4.5.2 Using fuzzy p-value

It is known [9] that under the null hypothesis ($\mu = \mu_0$) H_0 is rejected at significance level γ from a random sample of observations, if $p \leq \gamma$ and not rejected if $p > \gamma$, where p the crisp p-value of the test, which according to (4):

(a) for the *one-sided test from the left* (alternative hypothesis $H_1 : \theta < \theta_0$), is

$$p = \Phi(z_0) \tag{67}$$

(b) for the *one-sided test from the right* (alternative hypothesis $H_1 : \mu > \mu_0$),

$$p = 1 - \Phi(z_0) \tag{68}$$

(c) for the *two-sided test* (alternative hypothesis $H_1 : \mu \neq \mu_0$),

$$p = \begin{cases} 2\Phi(z_0), & z_0 \leq 0 \\ 2(1 - \Phi(z_0)), & z_0 > 0 \end{cases} \tag{69}$$

In the fuzzy case, the test of H_0 is based on the fuzzy statistic (61), the α-cuts of which for the given sample are given by (63). So, since the statistic Z follows the standard normal distribution, the α-cuts of the p-value:

(a) for the **one-sided test from the left,** according to (54) is

$$\begin{aligned} \overline{P}[\alpha] &= [\Pr(Z \leq z_0 - z_{h(\alpha)}), \Pr(Z \leq z_0 + z_{h(\alpha)})] \\ &= [\Phi(z_0 - z_{h(\alpha)}), \Phi(z_0 + z_{h(\alpha)})] \end{aligned} \tag{70}$$

where $\Phi(z)$ the distribution function of the standard normal distribution and z_0 and $z_{h(\alpha)}$ are given by (59) and (12).

(b) for the **one-sided test from the right,** according to (55):

$$\begin{aligned} \overline{P}[\alpha] &= [\Pr(Z \geq z_0 - z_{h(\alpha)}), \Pr(Z \geq z_0 + z_{h(\alpha)})] \\ &= [1 - \Phi(z_0 + z_{h(\alpha)}), \quad 1 - \Phi(z_0 - z_{h(\alpha)})] \end{aligned} \tag{71}$$

(c) for the **two-sided test,** according to (56) is:

$$\begin{aligned} \overline{P}[\alpha] &= \begin{cases} [2\Pr(Z \leq z_0 - z_{h(\alpha)}), \min\{1, 2\Pr(Z \leq z_0 + z_{h(\alpha)})\}], & z_0 \leq 0 \\ [2(\Pr(Z \geq z_0 + z_{h(\alpha)})), \min 1, 2(\Pr(Z \geq z_0 - z_{h(\alpha)}))], & z_0 > 0 \end{cases} \\ &= \begin{cases} [2\Phi(z_0 - z_{h(\alpha)}), \min\{1, 2\Phi(z_0 + z_{h(\alpha)})\}], & z_0 \leq 0 \\ [2(1 - \Phi(z_0 + z_{h(\alpha)})), \min\{1, 2(1 - \Phi(z_0 - z_{h(\alpha)}))\}], & z_0 > 0 \end{cases} \end{aligned} \tag{72}$$

because the median of the standard normal distribution is zero and the core of $\overline{Z}_0[\alpha]$ is obtained by (63) for $\alpha = 1$ to be $z_0 - z_{h(1)} = z_0 + z_{h(1)} = z_0$, since according to (6) $h(1) = 0.5$, so $z_{h(1)} = \Phi^{-1}(1 - 0.5) = 0$.

Having the α-cuts of the fuzzy numbers $\overline{P}$ and $\overline{S}$, H_0 is rejected or not as described in Section 4.4.2.

Example 12. We carry out the hypotheses tests of Example 11 using fuzzy p-value.

The crisp p-value of the two-sided test is given by (69) for $z_0 = 0.6$ (evaluated in Example 11),

$$p = 2(1 - \Phi(0.6)) = 0.5485 > \gamma = 0.05$$

Therefore, H_0 is not rejected by the crisp test.

(a) Implementing (72) and (57) for the two-sided test, we get the results of Figure 16, where the core of $\overline{P}$ is at the right of $\overline{S}$ and their point of intersection has $y_0 = 0.21$, so the truth value of $\overline{P} > \overline{S}$ is $1 - 0.21 = 0.79$. Therefore, H_0 is accepted by the one-sided test from left (significance level $\gamma = 0.05$) with degree of acceptance 0.79.

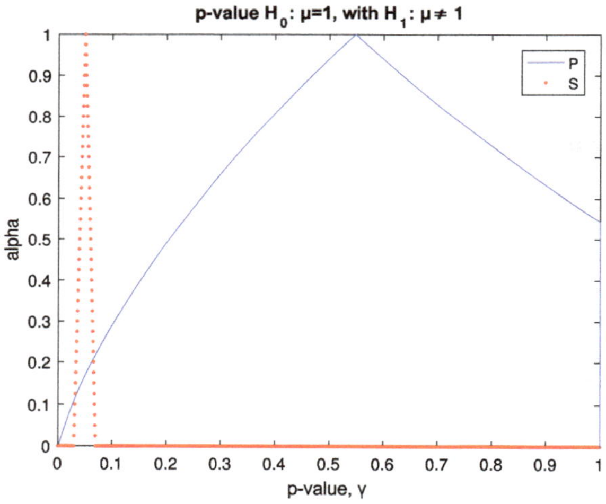

Figure 16: Fuzzy p-value $\overline{P}$ and significance level $\overline{S}$ for the two-sided test of Example 12.

(b) Implementing (69) and (57) for the one-sided test from the right, we get the results of Figure 17, where the core of $\overline{P}$ is at the right of $\overline{S}$ and their point of intersection has $y_0 = 0.3$, so the truth value of $\overline{P} > \overline{S}$ is $1 - 0.3 = 0.7$. Therefore, H_0 is accepted by the one-sided test from left (at significance level $\gamma = 0.05$) with degree of acceptance 0.7.

Example 13. We test at significance level 0.05 the null hypothesis $H_0 : \mu = 8$ with alternative $H_1 : \mu \neq 1$ (two-sided test) for the mean μ of a random variable X, which follows

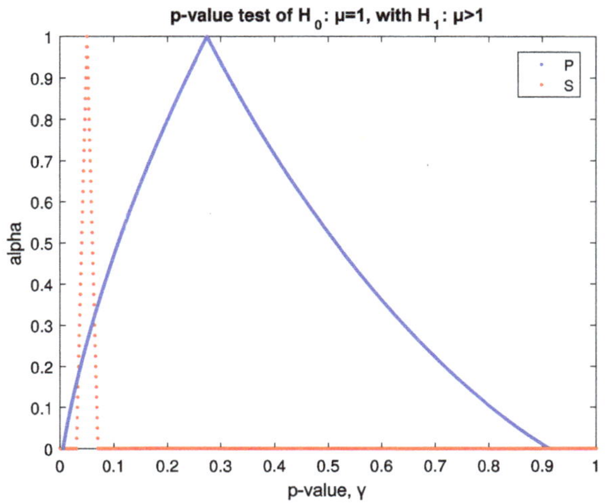

Figure 17: Fuzzy p-value $\overline{P}$ and significance level $\overline{S}$ for the one-sided test of Example 12.

normal distribution with variance $\sigma^2 = 4$ using a sample of $n = 30$ observations with sample mean $\overline{x} = 8$ with the fuzzy two-sided test:

(a) of Section 4.5.1 and

(b) of Section 4.5.2.

The value of the test statistic (58) is found by (59) to be

$$z_0 = \frac{8-8}{2/\sqrt{25}} = 0$$

Since $z_0 = 0 < z_{0.05/2} = 1.96$, H_0 is accepted by this crisp test.

Also, the crisp p-value of the test is given by (69),

$$p = 2\Phi(0) = 2 \cdot 0.5 = 1 > \gamma = 0.05,$$

so H_0 is accepted by this crisp test.

In this case, the crisp test gives no rejection of H_0 with the largest possible difference between the test statistic and the critical values, since the value of the test statistic is exactly in the middle of the no rejection region (or the p-value has its maximum value, 1), so it is the best case for acceptance of H_0.

(a) We apply the two-sided fuzzy test of H_0 implementing (63) and (66). So, we obtain the results of Figure 18, where the core of $\overline{Z}$ is between the cores of $\overline{\text{CV}}_1$ and $\overline{\text{CV}}_2$ and the point of intersection of $\overline{Z}$ and $\overline{\text{CV}}_1$ has $y_1 = 0.31$ and of $\overline{Z}$ and $\overline{\text{CV}}_2$, $y_2 = 0.31$. Hence,

$$\min((\overline{Z} > \overline{\text{CV}}_1), v(\overline{Z} < \overline{\text{CV}}_2)) = \min(1 - y_1, 1 - y_2) = 0.69,$$

so H_0 is accepted by this test with degree of acceptance $d = 0.69$.

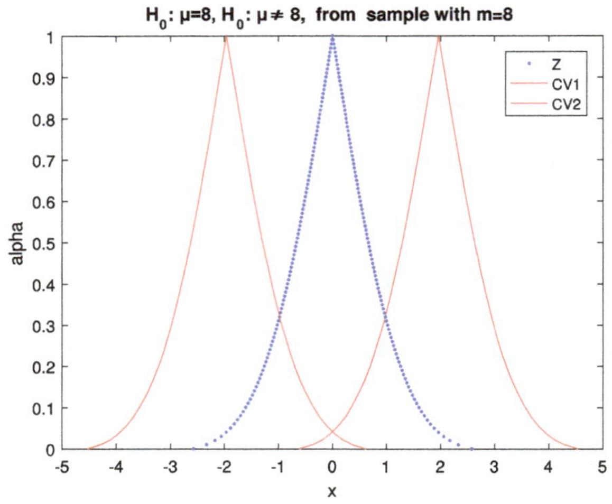

Figure 18: Fuzzy statistic $\overline{Z}$ and critical values $\overline{CV}_i$ for the two-sided test of Example 13.

(b) Implementing (72) and (57) for the two-sided test of H_0, we get the results of Figure 19, where the core of $\overline{P}$ is at the right of $\overline{S}$ and their point of intersection has $y_0 = 0.05$, so the truth value of $\overline{P} > \overline{S}$ is $1 - 0.05 = 0.95$. Therefore, H_0 is accepted by the one-sided test from left (at significance level $\gamma = 0.05$) with degree of acceptance 0.95.

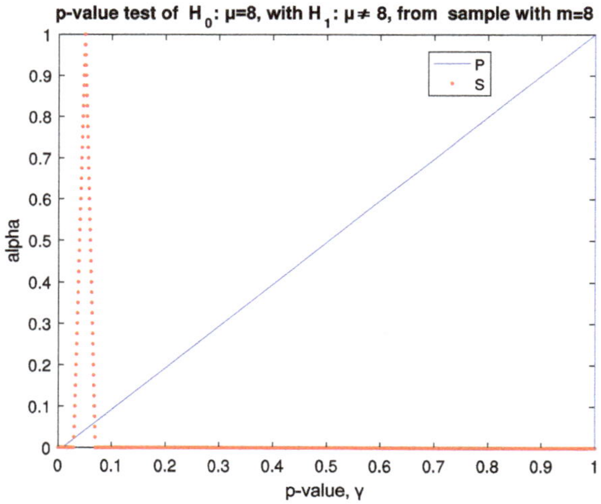

Figure 19: Fuzzy p-value $\overline{P}$ and significance level $\overline{S}$ for the two-sided test of Example 13.

4.6 Tests on the mean of a random variable with any distribution from a large sample

If the sample is large and X follows any distribution, then in the crisp case we test H_0 using the statistic

$$Z = \frac{\overline{X} - \mu_0}{s/\sqrt{n}}$$

where $\overline{X}$ is the statistic of the sample mean and s the sample standard deviation [9]. It is known that under the null hypothesis ($\mu = \mu_0$) Z follows the standard normal distribution $N(0,1)$ according to the central limit theorem [9], so H_0 is rejected or not from a large random sample of observations as in Sections 4.5.1 and 4.5.2, where the α-cuts of $\overline{Z}_0$, $\overline{\mathrm{CV}}_1$, $\overline{\mathrm{CV}}_2$, and $\overline{P}$ are given by (63)–(66) for the sample value

$$z_0 = \frac{\overline{x} - \mu_0}{s/\sqrt{n}} \tag{73}$$

of the statistic Z and the α-cuts of the fuzzy p-value by (70) and (71) for one-sided tests and by (72) for two-sided test.

Example 14. We test the null hypothesis $H_0 : \mu = 5$ at significance level $\gamma = 0.05$ with alternative $H_1 : \mu \neq 5$ (two-sided test) for the mean μ of a normal random variable X from two random samples of $n = 100$ observations of X with sample means and variances:

(a) $\overline{x} = 5.40$ and $s^2 = 4.62$;

(b) $\overline{x}_2 = 5.45$ and $s^2 = 4.69$

using fuzzy p-value.

(a) The value of the statistic Z for the first sample is found by (73) to be

$$z_0 = \frac{5.4 - 5}{\sqrt{4.62}/\sqrt{100}} = 1.86$$

so the crisp p-value of the test is found by (69) to be

$$p = 2(1 - \Phi(1.86)) = 0.063 > \gamma = 0.05.$$

Therefore, H_0 is not rejected by the crisp test for this sample.
Implementing (72) for the first sample ($\overline{x} = 5.4$) and fuzzifying the significance level to a triangular fuzzy number as in (57), we get the results of Figure 20, where the point of intersection of $\overline{P}$ and $\overline{S}$ has $y_0 = 0.93$. This according to (45) means that $v(\overline{P} \approx \overline{S}) = 0.93$, so we cannot make a decision on rejecting or not H_0 from this sample with any degree of confidence $d > 0.07$.

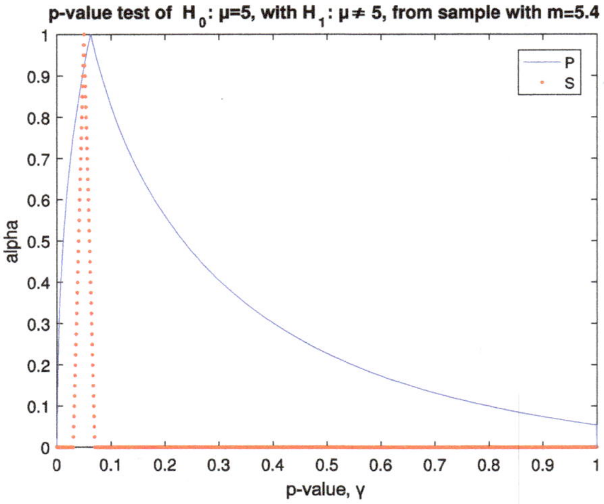

Figure 20: Fuzzy p-value $\overline{P}$ and significance level $\overline{S}$ for the two-sided test of $H_0 : \mu = 5$ of Example 14 from a sample with mean $\overline{x} = 5.4$.

(b) For the second sample, which has sample mean and variance $\overline{x}_2 = 5.45$ and $s^2 = 4.69$ (very close to these of the first sample), (73) gives

$$z_0 = \frac{5.45 - 5}{\sqrt{4.69}/\sqrt{100}} = 2.08,$$

so, the crisp p-value of the test is found by (69) to be

$$p = 2(1 - \Phi(2.08)) = 0.038 < \gamma = 0.05$$

Therefore, H_0 is rejected.

For the second sample ($\overline{x} = 5.45$), we get Figure 21, where the point of intersection of $\overline{P}$ and $\overline{S}$ has $y_0 = 0.94$. This according to (45) means that $v(\overline{P} \approx \overline{S}) = 0.91$, so we cannot make a decision on rejecting or not H_0 from this sample with any degree of confidence $d > 0.09$.

4.7 Tests on the means of two normal distributions with known variances

We test at significance level γ the null hypothesis

$$H_0 : \mu_1 = \mu_2$$

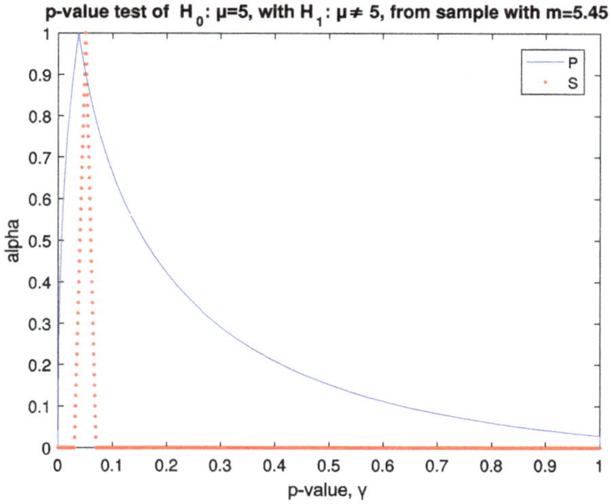

Figure 21: Fuzzy p-value $\overline{P}$ and significance level $\overline{S}$ for the two-sided test of $H_0 : \mu = 2$ of Example 14 from a sample with mean $\overline{x} = 5.45$.

for the means μ_1 and μ_2 of two random variables X_1 and X_2, which follow normal distributions with variances σ_1^2 and σ_2^2, using two independent random samples of n_1 and n_2 observations of X_1 and X_2 with sample means $\overline{x}_1$ and $\overline{x}_2$.

4.7.1 Using fuzzy critical values

It is known that under the null hypothesis (equal means) the test statistic

$$Z = \frac{\overline{X}_1 - \overline{X}_2}{\sigma_0}, \quad \text{where } \sigma_0 = \sqrt{\frac{\sigma_1^2}{n_1} + \frac{\sigma_2^2}{n_2}} \tag{74}$$

and $\overline{X}_1, \overline{X}_2$ the statistics of the sample means, follows standard normal distribution $N(0,1)$ [9]. So in the crisp case, H_0 is rejected from a given random sample:

(a) for the *one-sided test from the right*, if

$$z_0 > z_\gamma, \quad \text{where } z_\gamma = \Phi^{-1}(1-\gamma)$$

(b) for the *one-sided test from the left*, if $z_0 < -z_\gamma = \Phi^{-1}(\gamma)$,

(c) for the *two-sided* test, if $z_0 < -z_{\gamma/2}$ or $z_0 > z_{\gamma/2}$, where $z_{\gamma/2} = \Phi^{-1}(1-\gamma/2)$ and

$$z_0 = \frac{\overline{x}_1 - \overline{x}_2}{\sigma_0} \tag{75}$$

the value of the statistic Z for the given sample.

Otherwise, H_0 is not rejected.

For the fuzzy test of H_0, we use the fuzzy statistic [5]

$$\overline{Z} = \frac{\overline{\mu}_{12}}{\sigma_0} \tag{76}$$

which is generated by substituting $\overline{X}_1 - \overline{X}_2$ in (74) with the fuzzy estimator $\overline{\mu}_{12}$ of the difference of the sample means.

In our approach, we use the nonasymptotic fuzzy estimator $\overline{\mu}_{12}$ the α-cuts of which are (see Section 4.2.4)

$$\overline{\mu}_{12}[\alpha] = [\overline{x}_1 - \overline{x}_2 + z_{h(\alpha)}\sigma_0, \overline{x}_1 - \overline{x}_2 - z_{h(\alpha)}\sigma_0], \quad \alpha \in (0,1] \tag{77}$$

where $h(\alpha)$ is given by (6).

Using interval arithmetic from (74), (76) and (77) follows that the α-cuts of the fuzzy statistic $\overline{Z}$ are

$$\overline{Z}[\alpha] = [Z - z_{h(\alpha)}, Z + z_{h(\alpha)}], \quad \alpha \in (0,1] \tag{78}$$

For the given sample, we get the fuzzy number

$$\overline{Z}_0[\alpha] = [z_0 - z_{h(\alpha)}, z_0 + z_{h(\alpha)}], \quad \alpha \in (0,1] \tag{79}$$

where z_0 the sample value (75) of the statistic (74).

The fuzzy critical values are the fuzzy numbers $\overline{\text{CV}}_1$ and $\overline{\text{CV}}_2$, the α-cuts of which are given by (64), (65), or (66) for one- or two-sided tests for z_0 of (75). So, the rejection or acceptance of H_0 is done as in Section 4.4.1.

4.7.2 Using fuzzy *p*-value

According to (4), the crisp p-value of this test is given by (69) for the value z_0 of (75) of the statistic (74) for the given sample and according to (56) the α-cuts of the fuzzy p-value are given by (72) for z_0 of (75).

Having the α-cuts of the fuzzy numbers $\overline{P}$ and $\overline{S}$, H_0 is rejected or not as described in Section 4.4.2.

Example 15. We test at significance level 0.05 the null hypothesis $H_0 : \mu_1 = \mu_2$ with alternative the $H_1 : \mu_1 \neq \mu_2$ (two-sided test) for the means μ_1, μ_2 of the variables X_1 and X_2, which follow normal distributions with variances $\sigma_1^2 = 5.2$ and $\sigma_2^2 = 3.7$ from two independent random samples of observations of sizes $n_1 = 15$ and $n_2 = 8$ with sample means $\overline{x}_1 = 50.2$ and $\overline{x}_2 = 51.1$ using:

(a) fuzzy critical values,

(b) fuzzy p value.

(a) Implementing (63) and (66) for the fuzzy test of H_0 for z_0 evaluated by (75), we get the results of Figure 22, where the core of $\overline{Z}_0$ is between the cores of $\overline{\mathrm{CV}}_1$ and $\overline{\mathrm{CV}}_2$ and the point of intersection of $\overline{Z}_0$ and $\overline{\mathrm{CV}}_1$ has $y_1 = 0.6$ and of $\overline{Z}_0$ and $\overline{\mathrm{CV}}_2$, $y_2 = 0.1$, so

$$\min(\overline{Z}_0 > \overline{\mathrm{CV}}_1, v(\overline{Z}_0 < \overline{\mathrm{CV}}_2)) = \min(1 - y_1, 1 - y_2) = 0.4$$

Therefore, as described in Section 4.4.1, H_0 is accepted by this test with degree of acceptance $d = 0.4$.

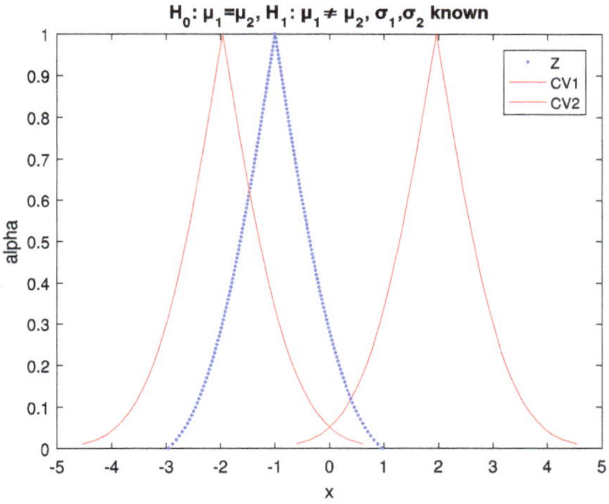

Figure 22: The fuzzy statistic $\overline{Z}$ and critical values for the test of Example 15.

(b) Implementing (72) and fuzzifying the significance level to a triangular fuzzy number as in (57), we get the results of Figure 23, where the core of $\overline{P}$ is at the right of the core of $\overline{S}$ and their point of intersection has $y_0 = 0.4$, so $v(\overline{P} > \overline{S}) = 1 - 0.4 = 0.6$. Therefore, H_0 is accepted by this test at significance level $y = 0.05$ with degree of acceptance 0.6.

4.8 Tests on the mean of a normal variable with unknown variance

We test at significance level y the null hypothesis

$$H_0 : \mu = \mu_0$$

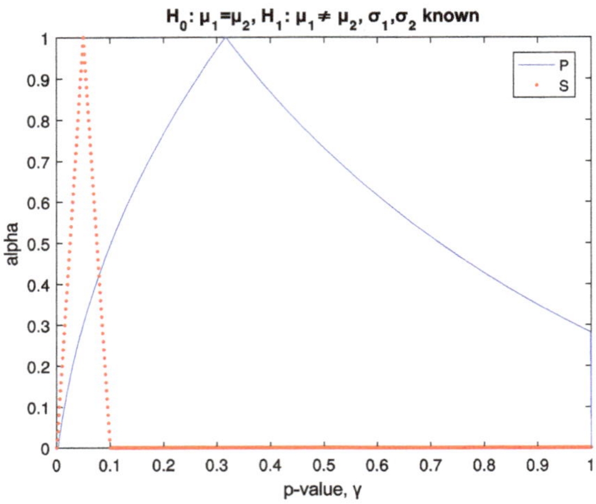

Figure 23: The fuzzy statistic $\overline{Z}$ and critical values for the test of Example 15.

for the mean μ of a random variable X, which follows normal distribution with unknown variance using a random sample of n observations of X with sample mean and variance $\overline{x}$ and s^2.

4.8.1 Using fuzzy critical values

In the crisp case, we test H_0 using the statistic

$$T = \frac{\overline{X} - \mu_0}{S/\sqrt{n}} \tag{80}$$

where $\overline{X}$ and S are the statistics of the sample mean and standard deviation.

It is known [9] that under the null hypothesis ($\mu = \mu_0$) T follows t-distribution with $n-1$ degrees of freedom. So in the crisp case, H_0 is rejected from a given random sample:

(a) for the *one-sided test from the left*, if $t_0 < -t_{\gamma;n-1}$,
(b) for the *one-sided test from the right*, if $t_0 > t_{\gamma;n-1}$,
(c) for the *two-sided* test, if $t_0 < -t_{\gamma/2;n-1}$ or $t_0 > t_{\gamma/2;n-1}$, where

$$t_0 = \frac{\overline{x} - \mu_0}{s/\sqrt{n}} \tag{81}$$

the value of the statistic (80),

$$t_{\gamma/2;n-1} = F_{n-1}^{-1}\left(1 - \frac{\gamma}{2}\right) \tag{82}$$

and F_{n-1}^{-1} the inverse distribution function of the t-distribution with $n-1$ degrees of freedom. While, if $-t_{\gamma/2;n-1} < t_0 < t_{\gamma/2;n-1}$, then H_0 is not rejected.

In the fuzzy case for the test of H_0, we use the fuzzy statistic

$$\overline{T} = \frac{\overline{\mu} - \mu_0}{\overline{\sigma}/\sqrt{n}} \tag{83}$$

which is generated by substituting $\overline{X}$ and S in (80) with the nonasymptotic fuzzy estimators $\overline{\mu}$ and $\overline{\sigma}$. Using (17), (38), (80), and fuzzy number arithmetic, the α-cuts of the fuzzy statistic $\overline{T}$ are found to be (omitting the implied n_1-1 degrees of freedom in χ^2 and t)

$$\begin{aligned}\overline{T}[\alpha] &= \left[\frac{\overline{X} - \mu_0 - t_{h(\alpha)}\frac{S}{\sqrt{n}}}{\frac{1}{\sqrt{n}}S\sqrt{\frac{(n-1)}{\chi^2_{L;h(\alpha)}}}}, \frac{\overline{X} - \mu_0 + t_{h(\alpha)}\frac{S}{\sqrt{n}}}{\frac{1}{\sqrt{n}}S\sqrt{\frac{(n-1)}{\chi^2_{R;h(\alpha)}}}}\right] \\ &= \left[\sqrt{\frac{\chi^2_{L;h(\alpha)}}{n-1}}(T - t_{h(\alpha)}), \sqrt{\frac{\chi^2_{R;h(\alpha)}}{n-1}}(T + t_{h(\alpha)})\right]\end{aligned}$$

where $t_{h(\alpha)}$ is given by (18). So, for the given sample we get the fuzzy number

$$\overline{T}_0[\alpha] = \left[\sqrt{\frac{\chi^2_{L;h(\alpha)}}{n-1}}(t_0 - t_{h(\alpha)}), \sqrt{\frac{\chi^2_{R;h(\alpha)}}{n-1}}(t_0 + t_{h(\alpha)})\right] \tag{84}$$

The fuzzy critical values of the test are the fuzzy numbers $\overline{\mathrm{CV}}_1$ and $\overline{\mathrm{CV}}_2$, the α-cuts of which are found as in Section 4.5.1 to be [5],

$$\begin{aligned}\overline{\mathrm{CV}}_1[\alpha] &= \left[\sqrt{\frac{\chi^2_{R;h(\alpha)}}{n-1}}(-t_{\gamma/2} - t_{h(\alpha)}), \sqrt{\frac{\chi^2_{L;h(\alpha)}}{n-1}}(-t_{\gamma/2} + t_{h(\alpha)})\right] \\ \overline{\mathrm{CV}}_2[\alpha] &= \left[\sqrt{\frac{\chi^2_{L;h(\alpha)}}{n-1}}(t_{\gamma/2} - t_{h(\alpha)}), \sqrt{\frac{\chi^2_{R;h(\alpha)}}{n-1}}(t_{\gamma/2} + t_{h(\alpha)})\right]\end{aligned} \tag{85}$$

The rejection or acceptance of H_0 is done as in Section 4.4.1 for $\overline{U}_0 = \overline{T}_0$.

4.8.2 Using fuzzy p-value

It is known [9] that under the null hypothesis ($\mu = \mu_0$) H_0 is rejected at significance level γ from a random sample of observations if $p \le \gamma$ and not rejected if $p > \gamma$, where p the crisp p-value of the test, which according to (4) is:

(a) for the *one-sided test from the left* (alternative hypothesis $H_1 : \theta < \theta_0$),

$$p = F_{n-1}(t_0) \tag{86}$$

(b) for the *one-sided test from the right* (alternative hypothesis $H_1 : \mu > \mu_0$),

$$p = 1 - F_{n-1}(t_0) \tag{87}$$

(c) for the *two-sided test* (alternative hypothesis $H_1 : \mu \neq \mu_0$),

$$p = \begin{cases} 2F_{n-1}(t_0), & t_0 \leq 0 \\ 2(1 - F_{n-1}(t_0)), & t_0 > 0 \end{cases} \tag{88}$$

where $F_{n-1}(t)$ the distribution function of the t_{n-1}-distribution and t_0 the value (81) of the test statistic (80).

The median of the t-distribution is equal to its mean, which is zero and the core of the $\overline{T}[\alpha]$ is obtained for $\alpha = 1$ to be

$$\sqrt{\frac{\chi^2_{L;0.5;n-1}}{n-1}}(t_0 - t_{h(1);n-1}) = \sqrt{\frac{m}{n-1}}t_0$$

where m the median of the χ^2_{n-1} distribution, since according to (6) $h(1) = 0.5$, so $t_{h(1);n-1} = t_{1-0.5;n-1} = 0$. Therefore, since the statistic T follows t_{n-1}-distribution, the α-cuts of the p-value of this test according to (56) are
if $\sqrt{\frac{m}{n-1}}t_0 \leq 0 \Leftrightarrow t_0 \leq 0$,

$$\begin{aligned} \overline{P}[\alpha] &= [2\Pr(T \leq \overline{T}_l[\alpha]), \min\{1, 2\Pr(T \leq \overline{T}_r[\alpha])\}] \\ &= \left[2\Pr\left(T \leq \sqrt{\frac{\chi^2_{L;h(\alpha)}}{n-1}}(t_0 - t_{h(\alpha)})\right),\right. \\ &\qquad \left.\min\left\{1, 2\Pr\left(T \leq \sqrt{\frac{\chi^2_{R;h(\alpha)}}{n-1}}(t_0 + t_{h(\alpha)})\right)\right\}\right], \end{aligned} \tag{89}$$

if $\sqrt{\frac{m}{n-1}}t_0 > 0 \Leftrightarrow t_0 > 0$,

$$\begin{aligned} \overline{P}[\alpha] &= [2\Pr(T \geq \overline{T}_l[\alpha]), \min\{1, 2\Pr(T \geq \overline{T}_r[\alpha])\}] \\ &= \left[2\Pr\left(T \geq \sqrt{\frac{\chi^2_{R;h(\alpha)}}{n-1}}(t_0 + t_{h(\alpha)})\right),\right. \\ &\qquad \left.\min\left\{1, 2\Pr\left(T \geq \sqrt{\frac{\chi^2_{L;h(\alpha)}}{n-1}}(t_0 - t_{h(\alpha)})\right)\right\}\right] \end{aligned} \tag{90}$$

Since $\overline{T}$ follows t_{n-1}-distribution (with $n - 1$ degrees of freedom), (89) and (90) give if $t_0 \leq 0$,

$$\overline{P}[\alpha] = \Bigg[2F_{n-1}\Bigg(\sqrt{\frac{\chi^2_{L;h(\alpha)}}{n-1}}(t_0 - t_{h(\alpha)})\Bigg), \min\Bigg\{1, 2F_{n-1}\Bigg(\sqrt{\frac{\chi^2_{R;h(\alpha)}}{n-1}}(t_0 + t_{h(\alpha)})\Bigg)\Bigg\}\Bigg] \quad (91)$$

if $t_0 > 0$,

$$\overline{P}[\alpha] = \Bigg[2\Bigg(1 - F_{n-1}\Bigg(\sqrt{\frac{\chi^2_{R;h(\alpha)}}{n-1}}(t_0 + t_{h(\alpha)})\Bigg)\Bigg), \min\Bigg\{1, 2\Bigg(1 - F_{n-1}\Bigg(\sqrt{\frac{\chi^2_{L;h(\alpha)}}{n-1}}(t_0 - t_{h(\alpha)})\Bigg)\Bigg)\Bigg\}\Bigg] \quad (92)$$

where $F_{n-1}(t)$ the distribution function of the t_{n-1}-distribution.

Having the α-cuts of the fuzzy numbers $\overline{P}$ and $\overline{S}$, H_0 is rejected or not as described in Section 4.4.2.

Example 16. We test the null hypothesis $H_0 : \mu = 1$ at significance level $\gamma = 0.05$ with alternative the (two-sided test) $H_1 : \mu \neq 1$ for the mean μ of a normal random variable X using a sample of $n = 25$ observations with sample mean and variance $\overline{x} = 1.32$ and $s^2 = 3.04$ using:
(a) fuzzy critical values;
(b) fuzzy p value.

In the crisp test, the sample value of the test statistic (80) is found by (81) to be

$$t_0 = \frac{1.32 - 1}{\sqrt{3.04}/\sqrt{25}} = 0.918$$

So, according to (88) the p-value of the crisp test is

$$p = 2(1 - F_{25-1}(0.918)) = 0.3678 > \gamma = 0.01$$

Therefore, H_0 is not rejected by this crisp test.

(a) Implementing (84) and (85) for the fuzzy test of H_0, we get the results of Figure 24, where the core of $\overline{Z}_0$ is between the cores of $\overline{\mathrm{CV}}_1$ and $\overline{\mathrm{CV}}_2$ and the point of intersection of $\overline{Z}_0$ and $\overline{\mathrm{CV}}_1$ has $y_1 = 0.12$ and of $\overline{Z}_0$ and $\overline{\mathrm{CV}}_2$, $y_2 = 0.42$, so

$$\min(\overline{Z}_0 > \overline{\mathrm{CV}}_1, v(\overline{Z}_0 < \overline{\mathrm{CV}}_2)) = \min(1 - y_1, 1 - y_2) = 0.58$$

Therefore, H_0 is accepted by this test with degree of acceptance $d = 0.58$.

(b) Implementing (91)–(92) and fuzzifying the significance level to a triangular fuzzy number as in (57), we get Figure 25, where the core of $\overline{P}$ is at the right the core of $\overline{S}$ and the point of intersection of $\overline{P}$ and $\overline{S}$ has $y_2 = 0.33$. So, H_0 is accepted by this test with degree of acceptance $d = 1 - 0.33 = 0.67$.

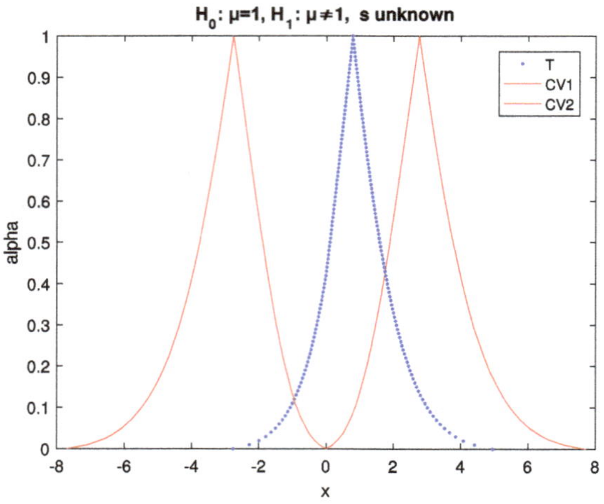

Figure 24: The fuzzy statistic $\overline{Z}$ and critical values for the two-sided test of $H_0 : \mu = 1$ of Example 16.

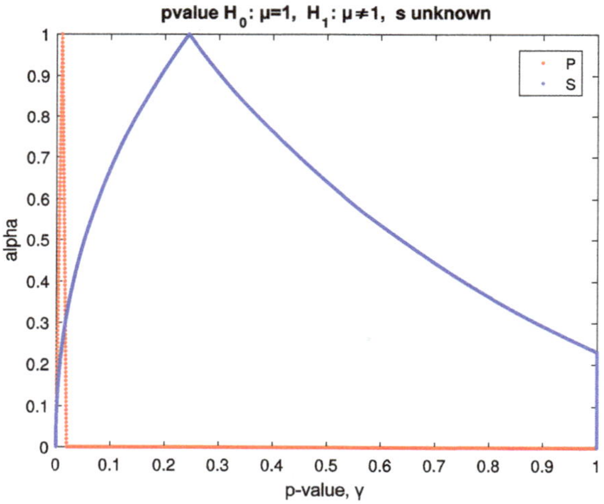

Figure 25: Fuzzy p-value $\overline{P}$ and significance level $\overline{S}$ for the two-sided test of $H_0 : \mu = 1$ of Example 16.

4.9 Tests on the variance of a normal distribution

We test at significance level γ the null hypothesis

$$H_0 : \sigma^2 = \sigma_0^2$$

for the variance σ^2 of a normal random variable X using a random sample of n observations of X and variance s^2.

4.9.1 Using fuzzy critical values

In the crisp case, we test H_0 using the test statistic

$$\chi^2 = \frac{(n-1)S^2}{\sigma_0^2} \tag{93}$$

(S the statistic of the sample variance) which follows χ^2_{n-1}-distribution with $n-1$ degrees of freedom [9]. So, H_0 is rejected from a given random sample (omitting in χ^2 the implied $n_1 - 1$ degrees of freedom):

(a) for the *one-sided test from the left*, if

$$\chi_0^2 < \chi^2_{L;\gamma}, \quad \text{where } \chi^2_{L;\gamma} = \chi^2_{1-\gamma} = F^{-1}(\gamma) \tag{94}$$

(b) for the *one-sided test from the right*, if

$$\chi_0^2 > \chi^2_{R;\gamma}, \quad \text{where } \chi^2_{R;\gamma} = \chi^2_{\gamma} = F^{-1}(1-\gamma) \tag{95}$$

and F^{-1}_{n-1} the inverse distribution function of the χ^2_{n-1}-distribution,

(c) for the *two-sided* test, if

$$\chi_0^2 < \chi^2_{L;\gamma/2} \quad \text{or} \quad \chi_0^2 > \chi^2_{R;\gamma/2}, \tag{96}$$

where

$$\chi^2_{L;\gamma/2} = \chi^2_{1-\gamma/2} = F^{-1}(\gamma/2) \quad \text{and} \quad \chi^2_{R;\gamma/2} = \chi^2_{\gamma/2} = F^{-1}(1-\gamma/2)$$

and

$$\chi_0^2 = \frac{(n-1)s^2}{\sigma_0^2} \tag{97}$$

the crisp value of the statistic (93). Otherwise, H_0 is not rejected.

In the fuzzy case for the test of H_0, we use the fuzzy statistic

$$\overline{\chi^2} = \frac{(n-1)\overline{\sigma^2}}{\sigma_0^2} \tag{98}$$

which is generated by substituting S^2 in (93) with a fuzzy estimator $\overline{\sigma^2}$ of the variance.

In our approach, we use the nonasymptotic fuzzy estimator $\overline{\sigma^2}$, the α-cuts of which are given by (37).

From (37), (93), (98), and interval arithmetic, we conclude that the α-cuts of the fuzzy statistic (98) are

$$\overline{\chi^2}[\alpha] = \left[\frac{(n-1)\chi^2}{\chi^2_{R;h(\alpha)}}, \frac{(n-1)\chi^2}{\chi^2_{L;h(\alpha)}}\right] \tag{99}$$

So, for the given sample we get from (99) the fuzzy number

$$\overline{\chi_0^2}[\alpha] = \left[\frac{(n-1)\chi_0^2}{\chi^2_{R;h(\alpha)}}, \frac{(n-1)\chi_0^2}{\chi^2_{L;h(\alpha)}}\right] \tag{100}$$

where χ_0^2 the crisp value (97) of the statistic (93) for the given sample.

Since the test statistic is fuzzy, the critical values are the fuzzy numbers $\overline{\mathrm{CV}}_1$ and $\overline{\mathrm{CV}}_2$, the α-cuts of which are defined and found as follows: ($\overline{\chi_l^2}[\alpha], \overline{\chi_r^2}[\alpha]$, $\overline{\mathrm{CV}}_{1l}[\alpha]$, $\overline{\mathrm{CV}}_{1r}[\alpha]$, and $\overline{\mathrm{CV}}_{2l}[\alpha]$, $\overline{\mathrm{CV}}_{2r}[\alpha]$ the left and right α-cuts of the test statistic $\overline{\chi^2}$ and of $\overline{\mathrm{CV}}_1$ and $\overline{\mathrm{CV}}_2$):

For the **two-sided test** (alternative hypothesis $H_1 : \sigma^2 \neq \sigma_0^2$) according to (52) and (99),

$$P(\overline{\chi_l^2}[\alpha] \geq \overline{\mathrm{CV}}_{2l}[\alpha]) = \frac{\gamma}{2} \Leftrightarrow P\left(\frac{(n-1)\chi^2}{\chi^2_{R;h(\alpha)}} \geq \overline{\mathrm{CV}}_{2l}[\alpha]\right) = \frac{\gamma}{2}$$
$$\Leftrightarrow P\left(\chi^2 \geq \frac{\chi^2_{R;h(\alpha)}\overline{\mathrm{CV}}_{2l}[\alpha]}{n-1}\right) = \frac{\gamma}{2}$$

So since the test statistic χ^2 follows χ^2_{n-1}-distribution,

$$\frac{\chi^2_{R;h(\alpha)}\overline{\mathrm{CV}}_{2l}[\alpha]}{n-1} = \chi^2_{R;\gamma/2} \Leftrightarrow \overline{\mathrm{CV}}_{2l}[\alpha] = \frac{(n-1)\chi^2_{R;\gamma/2}}{\chi^2_{R;h(\alpha)}}$$

Similarly, according to (53) and (99),

$$P(\overline{\chi_r^2}[\alpha] \geq \overline{\mathrm{CV}}_{2r}[\alpha]) = \frac{\gamma}{2} \Leftrightarrow \overline{\mathrm{CV}}_{2r}[\alpha] = \frac{(n-1)\chi^2_{R;\gamma/2}}{\chi^2_{L;h(\alpha)}}$$

Therefore, the α-cuts of $\overline{\mathrm{CV}}_2$ for the two-sided test are

$$\overline{\mathrm{CV}}_2[\alpha] = \left[\frac{(n-1)\chi^2_{R;\gamma/2}}{\chi^2_{R;h(\alpha)}}, \frac{(n-1)\chi^2_{R;\gamma/2}}{\chi^2_{L;h(\alpha)}}\right], \quad \alpha \in [0,1] \tag{101}$$

In the same way according to (50) and (99),

$$P(\overline{\chi_l^2}[\alpha] \leq \overline{\mathrm{CV}}_{1l}[\alpha]) = \frac{\gamma}{2} \Leftrightarrow P\left(\frac{(n-1)\chi^2}{\chi^2_{R;h(\alpha)}} \leq \overline{\mathrm{CV}}_{1l}[\alpha]\right) = \frac{\gamma}{2}$$
$$\Leftrightarrow P\left(\chi^2 \leq \frac{\chi^2_{R;h(\alpha)}\overline{\mathrm{CV}}_{1l}[\alpha]}{n-1}\right) = \frac{\gamma}{2}$$

So since the test statistic χ^2 follows χ^2_{n-1}-distribution,

$$\frac{\chi^2_{R;h(\alpha)}\overline{\mathrm{CV}}_{1l}[\alpha]}{n-1} = \chi^2_{L;\gamma/2} \Leftrightarrow \overline{\mathrm{CV}}_{1l}[\alpha] = \frac{(n-1)\chi^2_{L;\gamma/2}}{\chi^2_{R;h(\alpha)}}$$

Similarly, according to (51) and (99),

$$P(\overline{\chi_r^2}[\alpha] \leq \overline{\mathrm{CV}}_{1r}[\alpha]) = \frac{\gamma}{2} \Leftrightarrow \overline{\mathrm{CV}}_{1r}[\alpha] = \frac{(n-1)\chi^2_{L;\gamma/2}}{\chi^2_{L;h(\alpha)}}$$

Hence, the α-cuts of $\overline{\mathrm{CV}}_1$ for the two-sided test are

$$\overline{\mathrm{CV}}_1[\alpha] = \left[\frac{(n-1)\chi^2_{L;\gamma/2}}{\chi^2_{R;h(\alpha)}}, \frac{(n-1)\chi^2_{L;\gamma/2}}{\chi^2_{L;h(\alpha)}}\right], \quad \alpha \in [0,1] \tag{102}$$

Similarly, according to (46), (47), and (99), the α-cuts of $\overline{\mathrm{CV}}_1$ for the **one-sided test from the left** (alternative hypothesis $H_1 : \sigma^2 < \sigma_0^2$) are found to be

$$\overline{\mathrm{CV}}_1[\alpha] = \left[\frac{(n-1)\chi^2_{L;\gamma}}{\chi^2_{R;h(\alpha)}}, \frac{(n-1)\chi^2_{L;\gamma}}{\chi^2_{L;h(\alpha)}}\right], \quad \alpha \in [0,1] \tag{103}$$

and according to (48), (49), and (99), the α-cuts of $\overline{\mathrm{CV}}_2$ for the **one-sided test from the right** (alternative hypothesis $H_1 : \sigma^2 > \sigma_0^2$) are

$$\overline{\mathrm{CV}}_2[\alpha] = \left[\frac{(n-1)\chi^2_{R;\gamma}}{\chi^2_{R;h(\alpha)}}, \frac{(n-1)\chi^2_{R;\gamma}}{\chi^2_{L;h(\alpha)}}\right], \quad \alpha \in [0,1] \tag{104}$$

The rejection or acceptance of H_0 is done as in Section 4.4.1 for $\overline{U}_0 = \overline{\chi^2}_0$.

4.9.2 Using fuzzy *p*-value

The core of $\overline{\chi^2}$ is obtained by (100) for $\alpha = 1$ to be $(n-1)\chi_0^2/m$, where m the median of the χ^2_{n-1}-distribution, since from (6) $h(1) = 0.5$, so $\chi^2_{R;h(1);n-1} = \chi^2_{L;h(1);n-1} = \chi^2_{n-1}(0.5) = F^{-1}_{n-1}(0.5) = m$. Hence,

$$\frac{(n-1)\chi_0^2}{m} < m \Leftrightarrow \chi_0^2 < \frac{m^2}{n-1}$$

Therefore, according to (56) and (100), the α-cuts of the fuzzy p-value of the two-sided test of H_0 are

$$\overline{P}[\alpha] = \begin{cases} [2\Pr(\chi^2 \leq \frac{(n-1)\chi_0^2}{\chi^2_{R;h(\alpha)}}), \\ \quad \min\{1, 2\Pr(\chi^2 \leq \frac{(n-1)\chi_0^2}{\chi^2_{L;h(\alpha)}})\}], & \chi_0^2 \leq \frac{m^2}{n-1} \\ [2\Pr(\chi^2 \geq \frac{(n-1)\chi_0^2}{\chi^2_{L;h(\alpha)}}), \\ \quad \min\{1, 2\Pr(\chi^2 \geq \frac{(n-1)\chi_0^2}{\chi^2_{R;h(\alpha)}})\}], & \chi_0^2 > \frac{m^2}{n-1} \end{cases} \tag{105}$$

So since χ^2 follows χ^2_{n-1}-distribution with $n-1$ degrees of freedom, (105) gives if $\chi_0^2 \leq \frac{m^2}{n-1}$,

$$\overline{P}[\alpha] = \left[2F\left(\frac{(n-1)\chi_0^2}{\chi^2_{R;h(\alpha)}}\right), \min\left\{1, 2F\left(\frac{(n-1)\chi_0^2}{\chi^2_{L;h(\alpha)}}\right)\right\}\right], \tag{106}$$

if $\chi_0^2 > \frac{m^2}{n-1}$,

$$\overline{P}[\alpha] = \left[2\left(1-F\left(\frac{(n-1)\chi_0^2}{\chi^2_{L;h(\alpha)}}\right)\right), \min\left\{1, 2\left(1-F\left(\frac{(n-1)\chi_0^2}{\chi^2_{R;h(\alpha)}}\right)\right)\right\}\right] \tag{107}$$

where $F(x)$ the distribution function of the χ^2_{n-1}-distribution.

According to (54), the α-cuts of the fuzzy p-value of the one-sided test from the left are

$$\overline{P}[\alpha] = \left[F\left(\frac{(n-1)\chi_0^2}{\chi^2_{R;h(\alpha)}}\right), F\left(\frac{(n-1)\chi_0^2}{\chi^2_{L;h(\alpha)}}\right)\right]$$

and according to (55), the α-cuts of the one-sided test from the right

$$\overline{P}[\alpha] = \left[1-F\left(\frac{(n-1)\chi_0^2}{\chi^2_{L;h(\alpha)}}\right), 1-F\left(\frac{(n-1)\chi_0^2}{\chi^2_{R;h(\alpha)}}\right)\right]$$

Having the α-cuts of the fuzzy numbers $\overline{P}$ and $\overline{S}$, H_0 is rejected or not as described in Section 4.4.2.

Example 17. We test the null hypothesis $H_0 : \sigma^2 = 2$ at significance level $\gamma = 0.01$ with alternative the (two-sided test) $H_1 : \sigma^2 \neq 2$ for the variance σ^2 of a normal random variable X using a random sample of 101 observations of X with sample variance: (a) $s^2 = 2.77$, (b) $s^2 = 2.83$.

The crisp value of the test statistic (93) for the first sample is found by (97) to be

$$\chi_0^2 = \frac{(101-1)\cdot 2.77}{2} = 138.5$$

The critical values of this crisp test are (F^{-1}_{n-1} the inverse distribution function of the χ^2_{n-1}-distribution)

$$\chi^2_{L;0.01/2} = F^{-1}_{n-1}\left(\frac{\gamma}{2}\right) = 67.33 \quad \text{and} \quad \chi^2_{R;\gamma/2} = F^{-1}_{n-1}\left(1-\frac{0.01}{2}\right) = 140.17$$

So since $\chi^2_{L;0.01/2} < \chi_0^2 < \chi^2_{R;0.01/2}$, H_0 is not rejected by the crisp test for this sample.

For the second sample (variance $s^2 = 2.83$), the value of the test statistic (93) is found by (97) to be

$$\chi_0^2 = \frac{(101-1)\cdot 2.83}{2} = 141.5$$

So since $\chi_0^2 > \chi^2_{R;0.01/2}$, H_0 is rejected by the crisp test for this sample.

Implementing (100), (101), and (102) for the fuzzy test of H_0 for the first sample, we get the results of Figure 26, where the point of intersection of $\overline{\chi_0^2}$ and $\overline{\mathrm{CV}}_2$ has $y_2 = 0.98$. So according to (45), $v(\overline{\chi_0^2} \approx \overline{\mathrm{CV}}_2) = 0.98$. Therefore, we cannot make a decision on rejecting or not H_0 from this test.

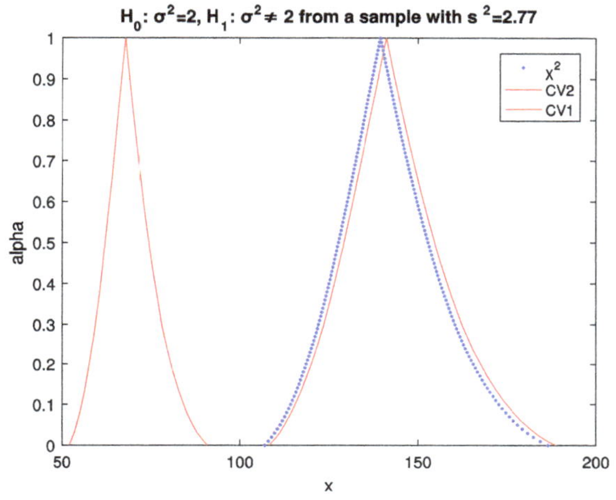

Figure 26: The fuzzy statistic $\overline{\chi}^2$ and the critical values for the test of Example 17 for a sample with variance $s^2 = 2.77$.

For the second sample, we get the results of Figure 27, where the point of intersection of the $\overline{\chi_0^2}$ and $\overline{\mathrm{CV}}_2$ has $y_2 = 0.99$. So according to (45), $v(\overline{\chi_0^2} \approx \overline{\mathrm{CV}}_2) = 0.99$. Hence, we cannot make a decision on rejecting or not H_0 from this test.

4.10 Tests on the variances of two normal distributions

We test at significance level γ the null hypothesis

$$H_0 : \sigma_1^2 = \sigma_2^2$$

for the variances σ_1^2 and σ_2^2 of two normal random variables X_1 and X_2 using two independent random samples of n_1 and n_2 observations of X_1 and X_2 with sample variances s_1^2 and s_2^2.

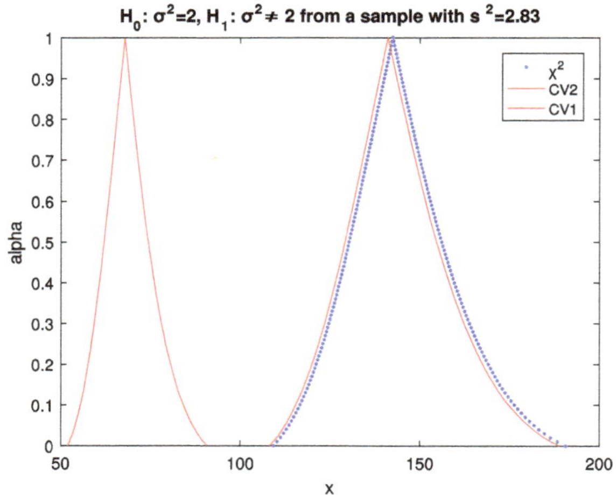

Figure 27: The fuzzy statistic $\overline{\chi}^2$ and the critical values for the test of Example 17 for a sample with variance $s^2 = 2.83$.

4.10.1 Using fuzzy critical values

It is known that

$$\chi_1^2 = \frac{(n_1 - 1)S_1^2}{\sigma_1^2} \sim \chi^2_{n_1-1} \quad \text{and} \quad \chi_2^2 = \frac{(n_2 - 1)S_2^2}{\sigma_2^2} \sim \chi^2_{n_2-1} \tag{108}$$

where S_1^2, S_2^2 the statistics of the sample variances [9]. So, in the crisp case under the null hypothesis (equal variances) the test statistic

$$F = \frac{\frac{\chi_1^2}{n_1-1}}{\frac{\chi_2^2}{n_2-1}} = \frac{S_1^2}{S_2^2} \tag{109}$$

follows a F_{n_1-1,n_2-1}-distribution with $n_1 - 1$ degrees of freedom for the numerator and $n_2 - 1$ degrees of freedom for the denominator [9]. Hence, H_0 is rejected from a given random sample:

(a) for the *one-sided test from the right*, if

$$f_0 \geq F_{1-\gamma/2} \tag{110}$$

(omitting in F_γ, $F_{\gamma/2}$, and $F_{1-\gamma/2}$ the implied $n_1 - 1, n_2 - 1$ degrees of freedom),

(b) for the *one-sided test from the left*, if

$$f_0 \leq F_\gamma \tag{111}$$

(c) for the *two-sided* test, if

$$f_0 \le F_{\gamma/2} \quad \text{or} \quad f_0 \ge F_{1-\gamma/2} \tag{112}$$

where ($F^{-1}_{\alpha;n_1-1,n_2-1}$ the inverse distribution function of the F_{n_1-1,n_2-1}-distribution)

$$F_{\gamma/2} = F^{-1}_{n_1-1,n_2-1}\left(1-\frac{\gamma}{2}\right) \quad \text{and} \quad F_{1-\gamma/2} = F^{-1}_{n_1-1,n_2-1}\left(\frac{\gamma}{2}\right) \tag{113}$$

and

$$f_0 = \frac{s_1^2}{s_2^2} \tag{114}$$

the crisp value of the statistic (109). Otherwise, H_0 is not rejected.

In the fuzzy test of H_0, we use the fuzzy statistic

$$\overline{F} = \frac{\overline{\sigma_1^2}}{\overline{\sigma_2^2}} \tag{115}$$

which is generated by substituting S_1^2, S_2^2 in (109) with the nonasymptotic fuzzy estimators $\overline{\sigma_1^2}$ and $\overline{\sigma_2^2}$ of [16], the α-cuts of which are given by (37).

Since $\overline{F}$ is the ratio of the fuzzy numbers $\overline{\sigma_1^2}$ and $\overline{\sigma_2^2}$, its α-cuts are found using (37) and interval arithmetic,

$$\begin{aligned}\overline{F}[\alpha] &= \left[\frac{\frac{(n_1-1)S_1^2}{\chi^2_{R;h(\alpha);n_1-1}}}{\frac{(n_2-1)S_2^2}{\chi^2_{L;h(\alpha);n_2-1}}}, \frac{\frac{(n_1-1)S_1^2}{\chi^2_{L;h(\alpha);n_1-1}}}{\frac{(n_2-1)S_2^2}{\chi^2_{R;h(\alpha);n_2-1}}}\right] \\ &= \left[\frac{(n_1-1)S_1^2\,\chi^2_{L;h(\alpha);n_2-1}}{(n_2-1)S_2^2\,\chi^2_{R;h(\alpha);n_1-1}}, \frac{(n_1-1)S_1^2\,\chi^2_{R;h(\alpha);n_2-1}}{(n_2-1)S_2^2\,\chi^2_{L;h(\alpha);n_1-1}}\right]\end{aligned} \tag{116}$$

For the two given random samples of X_1 and X_2 with sample variances s_1^2 and s_2^2, (116) gives

$$\overline{F}_0[\alpha] = \left[\frac{(n_1-1)s_1^2\,\chi^2_{L;h(\alpha);n_2-1}}{(n_2-1)s_2^2\,\chi^2_{R;h(\alpha);n_1-1}}, \frac{(n_1-1)s_1^2\,\chi^2_{R;h(\alpha);n_2-1}}{(n_2-1)s_2^2\,\chi^2_{L;h(\alpha);n_1-1}}\right] \tag{117}$$

Since the test statistic is fuzzy, the critical values are the fuzzy numbers $\overline{\mathrm{CV}}_1$ and $\overline{\mathrm{CV}}_2$, the α-cuts of which are defined and found according to (116) as follows ($\overline{F}_l[\alpha], \overline{F}_r[\alpha]$, $\overline{\mathrm{CV}}_{1l}[\alpha], \overline{\mathrm{CV}}_{1r}[\alpha]$, and $\overline{\mathrm{CV}}_{2l}[\alpha], \overline{\mathrm{CV}}_{2r}[\alpha]$ the left and right α-cuts of the test statistics $\overline{F}$, $\overline{\mathrm{CV}}_1$, and $\overline{\mathrm{CV}}_2$):

For the **two-sided test** (alternative hypothesis $H_1 : \sigma_1^2 \neq \sigma_2^2$) according to (52) and (116),

$$P(\overline{F}_l[\alpha] \geq \overline{\mathrm{CV}}_{2l}[\alpha]) = \frac{\gamma}{2} \Leftrightarrow P\left(\frac{(n_1-1)S_1^2}{(n_2-1)S_2^2}\frac{\chi^2_{L;h(a);n_2-1}}{\chi^2_{R;h(a);n_1-1}} \geq \overline{\mathrm{CV}}_{2l}[\alpha]\right) = \frac{\gamma}{2}$$

$$\Leftrightarrow P\left(\frac{S_1^2}{S_2^2} \geq \frac{(n_2-1)}{(n_1-1)}\frac{\chi^2_{R;h(a);n_1-1}}{\chi^2_{L;h(a);n_2-1}}\overline{\mathrm{CV}}_{2l}[\alpha]\right) = \frac{\gamma}{2}$$

$$\Leftrightarrow P\left(F \geq \frac{(n_2-1)}{(n_1-1)}\frac{\chi^2_{R;h(a);n_1-1}}{\chi^2_{L;h(a);n_2-1}}\overline{\mathrm{CV}}_{2l}[\alpha]\right) = \frac{\gamma}{2}$$

So, since the test statistic F follows F_{n_1-1,n_2-1}-distribution,

$$\frac{(n_2-1)}{(n_1-1)}\frac{\chi^2_{R;h(a);n_1-1}}{\chi^2_{L;h(a);n_2-1}}\overline{\mathrm{CV}}_{2l}[\alpha] = F_{\gamma/2}$$

Hence,

$$\overline{\mathrm{CV}}_{2l}[\alpha] = \frac{(n_1-1)\chi^2_{L;h(a);n_2-1}}{(n_2-1)\chi^2_{R;h(a);n_1-1}}F_{\gamma/2}$$

Similarly, according to (53) and (116),

$$P(\overline{F}_r[\alpha] \geq \overline{\mathrm{CV}}_{2r}[\alpha]) = \frac{\gamma}{2} \Leftrightarrow \overline{\mathrm{CV}}_{2r}[\alpha] = \frac{(n_1-1)\chi^2_{R;h(a);n_2-1}}{(n_2-1)\chi^2_{L;h(a);n_1-1}}F_{\gamma/2}$$

Therefore, the α-cuts of the fuzzy critical value $\overline{\mathrm{CV}}_2$ are

$$\overline{\mathrm{CV}}_2[\alpha] = \left[\frac{(n_1-1)\chi^2_{L;h(a);n_2-1}}{(n_2-1)\chi^2_{R;h(a);n_1-1}}F_{\gamma/2}, \quad \frac{(n_1-1)\chi^2_{R;h(a);n_2-1}}{(n_2-1)\chi^2_{L;h(a);n_1-1}}F_{\gamma/2}\right] \tag{118}$$

In the same way, from (50) and (116) we get

$$P(\overline{F}_l[\alpha] \leq \overline{\mathrm{CV}}_{1l}[\alpha]) = \frac{\gamma}{2} \Leftrightarrow P\left(\frac{(n_1-1)S_1^2}{(n_2-1)S_2^2}\frac{\chi^2_{R;h(a);n_2-1}}{\chi^2_{L;h(a);n_1-1}} \leq \overline{\mathrm{CV}}_{1l}[\alpha]\right) = \frac{\gamma}{2}$$

$$\Leftrightarrow P\left(\frac{S_1^2}{S_2^2} \leq \frac{(n_2-1)}{(n_1-1)}\frac{\chi^2_{L;h(a);n_1-1}}{\chi^2_{R;h(a);n_2-1}}\overline{\mathrm{CV}}_{1l}[\alpha]\right) = \frac{\gamma}{2}$$

$$\Leftrightarrow P\left(F \leq \frac{(n_2-1)}{(n_1-1)}\frac{\chi^2_{L;h(a);n_1-1}}{\chi^2_{R;h(a);n_2-1}}\overline{\mathrm{CV}}_{1l}[\alpha]\right) = \frac{\gamma}{2}$$

So, since the test statistic F follows F_{n_1-1,n_2-1}-distribution,

$$\frac{(n_2-1)}{(n_1-1)}\frac{\chi^2_{L;h(a);n_1-1}}{\chi^2_{R;h(a);n_2-1}}\overline{\mathrm{CV}}_{1l}[\alpha] = F_{1-\gamma/2}$$

Hence,

$$\overline{\mathrm{CV}}_1[\alpha] = \frac{(n_1-1)\chi^2_{R;h(\alpha);n_2-1}}{(n_2-1)\chi^2_{L;h(\alpha);n_1-1}} F_{1-\gamma/2}$$

Similarly, from (51) and (116),

$$P(\overline{F}_r[\alpha] \le \overline{\mathrm{CV}}_{1r}[\alpha]) = \frac{\gamma}{2} \Leftrightarrow \overline{\mathrm{CV}}_{1r}[\alpha] = \frac{(n_1-1)\chi^2_{L;h(\alpha);n_2-1}}{(n_2-1)\chi^2_{R;h(\alpha);n_1-1}} F_{1-\gamma/2}$$

Therefore, the α-cuts of the fuzzy critical value $\overline{\mathrm{CV}}_1$ are

$$\overline{\mathrm{CV}}_1[\alpha] = \left[\frac{(n_1-1)\chi^2_{L;h(\alpha);n_2-1}}{(n_2-1)\chi^2_{R;h(\alpha);n_1-1}} F_{1-\gamma/2}, \frac{(n_1-1)\chi^2_{R;h(\alpha);n_2-1}}{(n_2-1)\chi^2_{L;h(\alpha);n_1-1}} F_{1-\gamma/2} \right] \tag{119}$$

For the **one-sided test from the left** (alternative hypothesis $H_1 : \sigma_1^2 < \sigma_2^2$) the α-cuts of $\overline{\mathrm{CV}}_1$ according to (46), (47), and (116) are found to be

$$\overline{\mathrm{CV}}_1[\alpha] = \left[\frac{(n_1-1)\chi^2_{L;h(\alpha);n_2-1}}{(n_2-1)\chi^2_{R;h(\alpha);n_1-1}} F_{1-\gamma}, \frac{(n_1-1)\chi^2_{R;h(\alpha);n_2-1}}{(n_2-1)\chi^2_{L;h(\alpha);n_1-1}} F_{1-\gamma} \right] \tag{120}$$

For the **one-sided test from the right** (alternative hypothesis $H_1 : \sigma_1^2 > \sigma_2^2$) the α-cuts of $\overline{\mathrm{CV}}_2$ according to (48), (49), and (116) are found to be

$$\overline{\mathrm{CV}}_2[\alpha] = \left[\frac{(n_1-1)\chi^2_{L;h(\alpha);n_2-1}}{(n_2-1)\chi^2_{R;h(\alpha);n_1-1}} F_{\gamma}, \frac{(n_1-1)\chi^2_{R;h(\alpha);n_2-1}}{(n_2-1)\chi^2_{L;h(\alpha);n_1-1}} F_{\gamma} \right] \tag{121}$$

4.10.2 Using fuzzy *p*-value

The core of $\overline{F}_0$ is obtained by (117) for $\alpha = 1$ to be

$$\frac{(n_1-1)s_1^2\,\chi^2_{L;h(1);n_2-1}}{(n_2-1)s_2^2\,\chi^2_{R;h(1);n_1-1}} = \frac{(n_1-1)s_1^2 m_{n_2-1}}{(n_2-1)s_2^2 m_{n_1-1}}$$

since from (6) $h(1) = 0.5$, so

$$\frac{\chi^2_{L;h(1);n_2-1}}{\chi^2_{R;h(1);n_1-1}} = \frac{\chi^2_{L;0.5;n_2-1}}{\chi^2_{R;0.5;n_1-1}} = \frac{F^{-1}_{n_2-1}(0.5)}{F^{-1}_{n_1-1}(0.5)} = \frac{m_{n_2-1}}{m_{n_1-1}}$$

where m_{n_1-1}, m_{n_2-1} the medians of the $\chi^2_{n_1-1}$ and $\chi^2_{n_2-1}$-distributions and $F^{-1}_{n_1-1}, F^{-1}_{n_1-1}$ their inverse distribution function. Hence, if $m = F^{-1}_{n_1-1,n_2-1}(0.5)$ the median of the

F_{n_1-1,n_2-1}-distribution ($F^{-1}_{n_1-1,n_2-1}$ the inverse distribution function of the F_{n_1-1,n_2-1}-distribution),

$$\frac{(n_1-1)s_1^2 m_{n_2-1}}{(n_2-1)s_2^2 m_{n_1-1}} < m \Leftrightarrow \frac{s_1^2}{s_2^2} < \frac{(n_2-1)m_{n_1-1}}{(n_1-1)m_{n_2-1}} m$$

Therefore, according to (56) and (117), the α-cuts of the p-value of the **two-sided** test of H_0 are:
if $\frac{s_1^2}{s_2^2} \leq \frac{(n_2-1)m_{n_1-1}}{(n_1-1)m_{n_2-1}} m$

$$\overline{P}[\alpha] = \left[2\Pr\left(F \leq \frac{(n_1-1)s_1^2\, \chi^2_{L;h(\alpha);n_2-1}}{(n_2-1)s_2^2\, \chi^2_{R;h(\alpha);n_1-1}}\right), \right. \\ \left. \min\left\{1, 2\Pr\left(F \leq \frac{(n_1-1)s_1^2\, \chi^2_{R;h(\alpha);n_2-1}}{(n_2-1)s_2^2\, \chi^2_{L;h(\alpha);n_1-1}}\right)\right\}\right], \tag{122}$$

if $\frac{s_1^2}{s_2^2} > \frac{(n_2-1)m_{n_1-1}}{(n_1-1)m_{n_2-1}} m$

$$\overline{P}[\alpha] = \left[2\Pr\left(F \geq \frac{(n_1-1)s_1^2\, \chi^2_{R;h(\alpha);n_2-1}}{(n_2-1)s_2^2\, \chi^2_{L;h(\alpha);n_1-1}}\right), \right. \\ \left. \min\left\{1, 2\Pr\left(F \geq \frac{(n_1-1)s_1^2\, \chi^2_{L;h(\alpha);n_2-1}}{(n_2-1)s_2^2\, \chi^2_{R;h(\alpha);n_1-1}}\right)\right\}\right] \tag{123}$$

Since F follows the F_{n_1-1,n_2-1}-distribution, (122) and (123) give
if $\frac{s_1^2}{s_2^2} \leq \frac{(n_2-1)m_{n_1-1}}{(n_1-1)m_{n_2-1}} m$, then

$$\overline{P}[\alpha] = \left[2F_{n_1-1,n_2-1}\left(\frac{(n_1-1)s_1^2\, \chi^2_{L;h(\alpha);n_2-1}}{(n_2-1)s_2^2\, \chi^2_{R;h(\alpha);n_1-1}}\right), \right. \\ \left. \min\left\{1, 2F_{n_1-1,n_2-1}\left(\frac{(n_1-1)s_1^2\, \chi^2_{R;h(\alpha);n_2-1}}{(n_2-1)s_2^2\, \chi^2_{L;h(\alpha);n_1-1}}\right)\right\}\right], \tag{124}$$

if $\frac{s_1^2}{s_2^2} > \frac{(n_2-1)m_{n_1-1}}{(n_1-1)m_{n_2-1}} m$, then

$$\overline{P}[\alpha] = \left[2\left(1 - F_{n_1-1,n_2-1}\left(\frac{(n_1-1)s_1^2\, \chi^2_{R;h(\alpha);n_2-1}}{(n_2-1)s_2^2\, \chi^2_{L;h(\alpha);n_1-1}}\right)\right), \right. \\ \left. \min\left\{1, 2\left(1 - F_{n_1-1,n_2-1}\left(\frac{(n_1-1)s_1^2\, \chi^2_{L;h(\alpha);n_2-1}}{(n_2-1)s_2^2\, \chi^2_{R;h(\alpha);n_1-1}}\right)\right)\right\}\right] \tag{125}$$

Similarly, according to (54) and (117), the α-cuts of the fuzzy p-value of the **one-sided test from the left** are

$$\overline{P}[\alpha] = \left[F_{n_1-1,n_2-1}\left(\frac{(n_1-1)s_1^2\,\chi^2_{L;h(\alpha);n_2-1}}{(n_2-1)s_2^2\,\chi^2_{R;h(\alpha);n_1-1}}\right),\right.$$
$$\left.F_{n_1-1,n_2-1}\left(\frac{(n_1-1)s_1^2\,\chi^2_{R;h(\alpha);n_2-1}}{(n_2-1)s_2^2\,\chi^2_{L;h(\alpha);n_1-1}}\right)\right] \tag{126}$$

and according to (55) and (117), of the **one-sided test from the right**

$$\overline{P}[\alpha] = \left[1 - F_{n_1-1,n_2-1}\left(\frac{(n_1-1)s_1^2\,\chi^2_{R;h(\alpha);n_2-1}}{(n_2-1)s_2^2\,\chi^2_{L;h(\alpha);n_1-1}}\right),\right.$$
$$\left.1 - F_{n_1-1,n_2-1}\left(\frac{(n_1-1)s_1^2\,\chi^2_{L;h(\alpha);n_2-1}}{(n_2-1)s_2^2\,\chi^2_{R;h(\alpha);n_1-1}}\right)\right] \tag{127}$$

Example 18.

(a) We test at significance level 0.05 the hypothesis $H_0 : \sigma_1^2 = \sigma_2^2$ using fuzzy critical values from two random samples of $n_1 = 11$ and $n_2 = 13$ observations with sample variances $s_1^2 = 0.24$ and $s_2^2 = 1.05$ with alternative:
  (i) $H_1 : \sigma_1^2 \neq \sigma_2^2$ (two-sided test);
  (ii) $H_1 : \sigma_1^2 < \sigma_2^2$ (one-sided test).

(b) We apply the test of a(*i*) using fuzzy p-value.

(a) (i) We apply the two-sided test of H_0 of Section 4.10.1 implementing (117), (118) and (119). So, we get the results of Figure 28, where the core of $\overline{Z}$ is between the cores of $\overline{\mathrm{CV}}_1$ and $\overline{\mathrm{CV}}_2$ and the point of intersection of $\overline{Z}$ and $\overline{\mathrm{CV}}_1$ has $y_1 = 0.84$ and the point of intersection of $\overline{Z}$ and $\overline{\mathrm{CV}}_2$ (not shown), $y_2 = 0.23$, so that

$$\min(\overline{Z}_0 > \overline{\mathrm{CV}}_1, v(\overline{Z}_0 < \overline{\mathrm{CV}}_2)) = \min(1 - y_1, 1 - y_2) = 0.16$$

Therefore, as described in Section 4.4.1, H_0 is accepted by this test with degree of acceptance $d = 0.16$.

(ii) Applying the above one-sided test from the left of H_0 implementing (117) and (120), we get the results of Figure 29, where the point of intersection of $\overline{Z}$ and $\overline{\mathrm{CV}}_1$ has $y_1 = 0.95$, so that

$$v(\overline{Z} \approx \overline{\mathrm{CV}}_1) = 0.95$$

Therefore, according to Section 4.4.1, we cannot make a decision on the rejection or not of H_0 by this test with any degree of confidence $d > 0.05$.

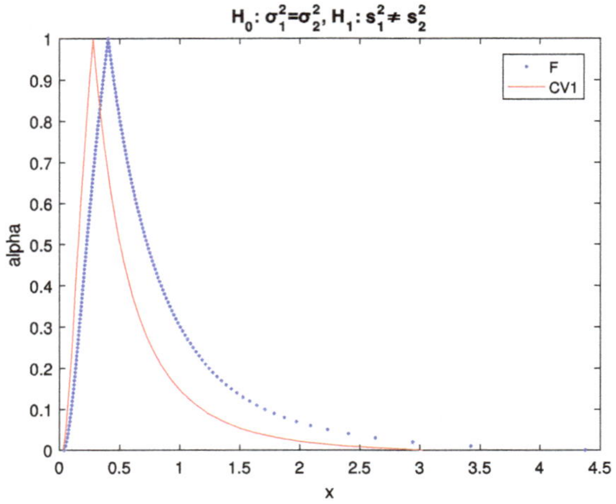

Figure 28: The fuzzy statistic $\overline{F}$ and critical value $\overline{CV}_1$ for the two-sided test of Example 18a.

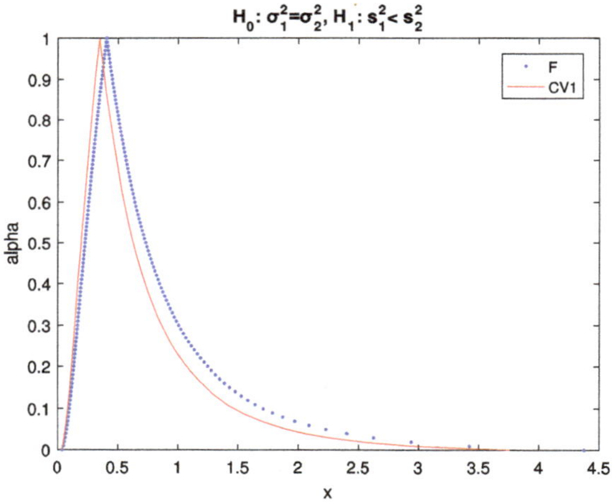

Figure 29: The fuzzy statistic $\overline{F}$ and critical value $\overline{CV}_1$ for the one-sided test of Example 18b.

(b) Applying the two-sided test of H_0 of Section 4.10.2 implementing (124), (125), and (57), we get the results of Figure 30, where the core of $\overline{P}$ is at the right of the core of $\overline{S}$ and their point of intersection has $y_0 = 0.79$, so that

$$v(\overline{P} > \overline{S}) = 1 - 0.79 = 0.21$$

Therefore, as described in Section 4.4.2, H_0 is accepted by this test with degree of acceptance $d = 0.21$.

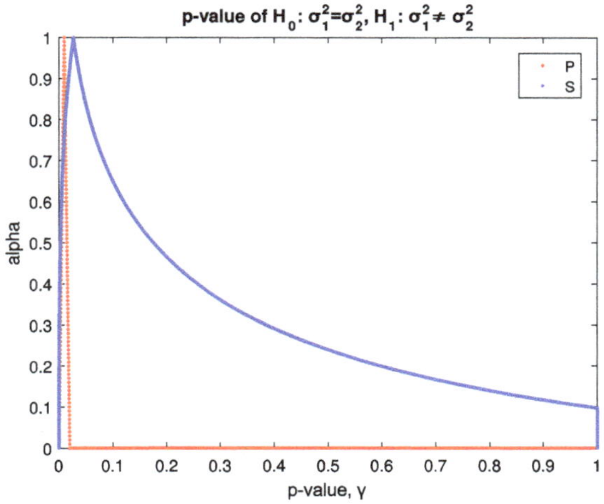

Figure 30: Fuzzy p-value $\overline{P}$ and significance level $\overline{S}$ for the two-sided test of $H_0 \ : \ \mu \ = \ 1$ of Example 18b.

4.11 Conclusions

If the value of the test statistic of a hypothesis is close to a critical value of the test, then the crisp hypothesis test is unstable, since a very small change in the sample may lead from rejection to no rejection of H_0 or vice versa, as in Examples 14 and 17. Whereas our approach, which uses nonasymptotic fuzzy estimators for the construction of the test statistic and a rejection or acceptance degree of the null hypothesis H_0, gives better results than the crisp test, since it gives us the possibility to make a decision on a partial rejection or acceptance of H_0, as shown in Examples 11b, 14, and 17 where we get a very low degree of rejection or acceptance of the null hypothesis, which means "no decision."

In the crisp statistics, the hypotheses tests, which use critical values, are equivalent to the tests, which use p-value. This does not happen in the fuzzy hypotheses tests, since in general the tests which use fuzzy critical values give different degrees of rejection or acceptance of H_0 than the tests, which use fuzzy p-value. The two tests give similar results only in cases in which the p-value of the test is close to the significance level (or equivalently if the test statistic is close to a critical value), as in Example 18, but in the other cases the former gives significantly lower degree of rejection or acceptance than the later (Examples 11a–12, 13, and 15). Characteristic is the case of the hypotheses tests of Example 13, in which the crisp p-value is one or very close to 1 (the value of the test statistic is in the center of the acceptance region), which is the best case of acceptance. As shown in this example, the test which uses fuzzy critical values gives significantly lower degree of acceptance of H_0 than the expected, which is a value close to one (d = 0.68), whereas the test which uses fuzzy p-value gives degree of acceptance of H_0 near to one ($d = 0.95$). So since the test which uses fuzzy p-value is in much better agreement with the respective crisp test than the test which uses fuzzy critical values, we can conclude that the former gives more reliable results than the later.

Bibliography

[1] J. J. Buckley, On the algebra of interactive fuzzy numbers, Fuzzy Sets Syst. 32 (1989), 291–306.

[2] J. J. Buckley, Fuzzy probabilities: new approach and applications, Springer, Berlin, Heidelberg, 2005.

[3] J. J. Buckley, Uncertain probabilities II: the continuous case, Soft Comput. 8 (2004), 193–199.

[4] J. J. Buckley, Fuzzy statistics: hypotheses testing, Soft Comput. 9 (2005), 512–518.

[5] J. J. Bucley, Fuzzy statistics, Springer, Berlin, Heidelberg, 2004.

[6] D. Dubois and H. Prade, Ranking of fuzzy numbers in the setting of possibility theory, Inf. Sci. 30 (1983), 183–224.

[7] D. Dubois and H. Prade, Fuzzy Sets and Systems: Theory and applications, Academic Press, 1980.

[8] P. Filzmoser and R. Viertl, Testing hypotheses with fuzzy data: The fuzzy p-value, Metrika 59 (2004), 21–29.

[9] R. V. Hogg and E. A. Tanis, Probability and Statistical Inference, 6th edn, Prentice Hall, Upper Saddle River, N. J., 2001.

[10] M. Holena, Fuzzy hypotheses testing in a framework of fuzzy logic, Fuzzy Sets Syst. 145 (2004), 229–252.

[11] H. Lee and J. H. Lee, A method for ranking fuzzy numbers and its application to a decision-making, IEEE Trans. Fuzzy Syst. 7(6) (1999), 677–685.

[12] N. Mylonas and B. Papadopoulos, Hypotheses tests using non-asymptotic fuzzy estimators and fuzzy critical values, in: Proceedings of the 16th Conference on Artificial Intelligence and Innovations, 2020.

[13] N. Mylonas and B. Papadopoulos, Fuzzy hypotheses tests for crisp data using non-asymptotic fuzzy estimators and a degree of rejection or acceptance, (preprint).

[14] N. Mylonas and B. Papadopoulos, Fuzzy p-value of hypotheses tests with crisp data using non-asymptotic fuzzy estimators, J. Stoch. Anal. 2(1) (2021), 1. http://dx.doi.org/10.31390/josa.2.1.01

[15] A. Parchami and S. Mahmoud Taheri, Fuzzy p-value in testing fuzzy hypotheses with crisp data, Stat. Pap. 51 (2010), 209–226.

[16] D. Sfiris and B. Papadopoulos, Non-asymptotic fuzzy estimators based on confidence intervals, Inf. Sci. 279 (2014), 446–459.

[17] M. Taheri and M. Arefi, Testing fuzzy hypotheses based on fuzzy test statistic, Soft Comput. 13 (2009), 617–625.

[18] M. Taheri and G. Hesamian, Non-parametric Statistical Tests for Fuzzy Observations. Fuzzy Test Statistic Approach, Int. J. Fuzzy Log. Intell. Syst. 17(3) (2017), 145–153.

[19] R. Viertl, Statistical Methods for Fuzzy Data, Wiley, 2011.

[20] L. A. Zadeh, Fuzzy sets, Inf. Control 8(3) (1965), 338–353.

Jarosław Pykacz

5 Fuzzy sets in quantum mechanics

Abstract: Summary of more than 30 years lasting author's attempts at utilizing fuzzy set ideas in quantum mechanics is given. Similar attempts of some other researchers are briefly mentioned.

5.1 Introduction

The possibility of using fuzzy sets in the description of quantum phenomena is yielded by the following observations:

1. Membership functions of fuzzy sets take values in the interval $[0, 1]$, and often these values can be interpreted as degrees to which objects possess their properties. Probabilities, extensively used in quantum mechanics, also take values in the interval $[0, 1]$. It should be stressed, that probabilities in quantum mechanics are not "epistemic" but "ontic," that is, they are not caused by our lack of knowledge of exact values of parameters that characterize quantum objects, but by "intrinsic" properties of these objects.
2. Families of closed linear subspaces of Hilbert spaces that describe properties of quantum objects are isomorphic with some specific families of fuzzy subsets of sets of states of these objects.

The first observation means that, for example, instead of saying: "Probability that a linearly polarized photon will pass through a linear polarizer oriented under the angle α, is $\cos^2\alpha$' (Malus law)," one can say: "A linearly polarized photon possesses the property of being able to pass through a linear polarizer oriented under the angle α to the degree $\cos^2\alpha$', and of course this photon at the same time possesses the property of NOT being able to pass through this polarizer to the degree $1 - \cos^2\alpha = \sin^2\alpha$."

This is of course in a strong disagreement with the "classical" point of view, based on the classical two-valued logic, according to which every physical object either entirely possesses, or does not possess any of its properties.

Experimentally confirmed violation of the so-called *Bell inequalities* (see, e. g., [1, 13]), the derivation of which is based on the assumption that all physical objects either entirely possess or entirely do not possess each of their properties, also indicate that description of quantum phenomena based on fuzzy sets (or, equivalently, many-valued logic) is more adequate than description based on classical two-valued logic and traditional "crisp" sets. It should be also mentioned that in fuzzy probability calculus Bell-type inequalities can be violated even in the case of classical, macroscopic objects [36], and that fuzzy probabilities can replace much less plausible nega-

https://doi.org/10.1515/9783110704303-005

tive probabilities in attempts to explain violation of Bell-type inequalities by quantum objects [32].

Explanation of the second observation requires introducing some mathematical notions and will be presented in the next section.

5.2 Fuzzy set representation of Birkhof–von Neumann "quantum logic"

After Garrett Birkhoff and John von Neumann published their 1936 paper [4], it is generally accepted that a family of all properties of a quantum system is mathematically represented by a family of all closed linear subspaces $L(\mathcal{H})$ of a Hilbert space $\mathcal{H}$ used to describe this system. From the algebraic point of view, such families are orthomodular lattices, that is, a nondistributive generalization of Boolean algebras (we remind that Boolean algebras represent families of properties of classical physical systems). Orthomodular lattices, or their slight generalizations – orthomodular partially ordered sets are usually called "quantum logics."

Definition 1. By a *quantum logic,* we mean an orthocomplemented σ-orthocomplete orthomodular poset, that is, a partially ordered set L, which contains the smallest element O and the greatest element I, in which the orthocomplementation map $\perp: L \to L$ satisfying the conditions (a)–(c) exists:
(a) $(a^\perp)^\perp = a$.
(b) If $a \leq b$, then $b^\perp \leq a^\perp$.
(c) The greatest lower bound (*meet*) $a \wedge a^\perp$ and the least upper bound (*join*) $a \vee a^\perp$ with respect to the given partial order exist in L and $a \wedge a^\perp = O$, $a \vee a^\perp = I$.

Moreover, the σ-orthocompleteness condition holds:
(d) If $a_i \leq a_j^\perp$ for $i \neq j$ (such elements are called *orthogonal* and are usually denoted $a_i \perp a_j$), then the join $\vee_i a_i$ exists in L,

and so does the orthomodular identity:
(e) If $a \leq b$, then $b = a \vee (a^\perp \wedge b) = a \vee (a \vee b^\perp)^\perp$.

Two elements $a, b \in L$ are called *compatible* iff there exist in L three pairwisely orthogonal elements a_1, b_1, c such that $a = a_1 \vee c$ and $b = b_1 \vee c$. It occurs that in the case of an orthomodular lattice $L(\mathcal{H})$ consisting of all closed subspaces of a Hilbert space $\mathcal{H}$; two subspaces $\mathbf{A}, \mathbf{B} \in L(\mathcal{H})$ are compatible iff operators of orthogonal projections $\hat{A}$ and $\hat{B}$ onto these subspaces commute.

In Boolean algebras, all elements are compatible

Definition 2. *Probability measure* on a quantum logic L is a mapping $s : L \to [0,1]$ such that:

(i) $s(I) = 1$;
(ii) $s(\vee_i a_i) = \Sigma_i s(a_i)$ for any sequence of pairwise orthogonal elements of L.

If elements of a quantum logic L represent properties of a physical system, then probability measures defined on L represent states of this system and, therefore, are often themselves called *states* on L. According to the standard interpretation, a number $s(a) \in [0,1]$ is interpreted as a probability of obtaining the positive result in an experiment designed to check whether a physical system has a property represented by a when this system is in a state represented by s. However, as we already mentioned, this number can also be interpreted as a degree to which a physical system in a state represented by s has a property represented by a.

A set of probability measures (states) S on a quantum logic L is called *ordering* (*full, order determining*) iff $s(a) \leq s(b)$ for all $s \in S$ implies $a \leq b$. Let us note that the only way in which one can establish experimentally partial order relation between various elements of L is to conduct experiments on a system prepared in various states, which means that only quantum logics that allow ordering sets of probability measures can be endowed with physical interpretation. Therefore, throughout the rest of this chapter we shall consider only quantum logics with ordering sets of probability measures.

An interested reader is referred to any textbook on quantum logic theory (see, e. g., [3] or [25]) for precise mathematical definitions and physical interpretation of all these notions.

The present author showed in a series of papers [26–34] (the main results collected in a book [35]), that any physically sound quantum logic in a Birkhoff–von Neumann sense can be isomorphically represented as a family of fuzzy subsets of the set of states of a physical system under study. This family has to fulfill some specific conditions but before we list them we have to remind some notions.

Definition 3. *Łukasiewicz union* and *Łukasiewicz intersection*[1] of fuzzy sets are defined, respectively, as follows:

$$(A \sqcup B)(s) = \min[A(s) + B(s), 1], \tag{1}$$
$$(A \sqcap B)(s) = \max[A(s) + B(s) - 1, 0]. \tag{2}$$

Two fuzzy sets such that

$$A \sqcap B = \emptyset \tag{3}$$

were called *weakly disjoint* in [11].

1 Robin Giles was the first who studied these operations within the area of fuzzy sets in [11]. He denoted them by bold symbols; therefore, they are also called *bold*, *Giles*, *truncated*, *bounded*, or *arithmetic* operations. However, nowadays the name *Łukasiewicz operations* is the most commonly used. Etymology of this name can be explained by close links between Łukasiewicz many-valued logics and fuzzy set theory (see, e. g., [11] or [35]).

The following theorem was proved in [29].

Theorem 1. *Any quantum logic L with an ordering set of probability measures S can be isomorphically represented in the form of a family $\mathcal{L}(S)$ of fuzzy subsets of S satisfying the following conditions:*
(a) *$\mathcal{L}(S)$ contains the empty set $\emptyset$.*
(b) *$\mathcal{L}(S)$ is closed with respect to the standard fuzzy set complementation, that is, if $A \in \mathcal{L}(S)$, then $A' \in \mathcal{L}(S)$.*
(c) *$\mathcal{L}(S)$ is closed with respect to countable Łukasiewicz unions of pairwise weakly disjoint sets. In symbols: if $A_i, A_j \in \mathcal{L}(S)$, $A_i \sqcap A_j = \emptyset$ for $i \neq j$, then $\sqcup_i A_i \in \mathcal{L}(S)$.*
(d) *The empty set $\emptyset$ is the only set in $\mathcal{L}(S)$ that is weakly disjoint with itself, that is, for any $A \in \mathcal{L}(S)$, if $A \sqcap A = \emptyset$, then $A = \emptyset$.*

Conversely, any family of fuzzy subsets of an arbitrary universe $\mathcal{U}$ satisfying conditions (a)–(d) is a quantum logic partially ordered by inclusion of fuzzy sets, with the standard fuzzy set complementation as orthocomplementation, orthogonality of elements coinciding with their weak disjointness, and an ordering set of probability measures generated by points of the universe $\mathcal{U}$ according to the formula

$$s_x(A) = A(x) \quad \text{for all } x \in \mathcal{U}. \tag{4}$$

Because of the second part of this theorem, any family $\mathcal{L}(\mathcal{U})$ of fuzzy subsets of an arbitrary universe $\mathcal{U}$ that satisfies conditions (a)–(d) of Theorem 1 is called a *quantum logic of fuzzy sets* or simply a *fuzzy quantum logic*. Let us note that Łukasiewicz operations in a fuzzy quantum logic are, in general, defined only partially since it was proved in [31] that if they are globally defined, a fuzzy quantum logic is necessarily a Boolean algebra.

If we assume, as it is done in quantum mechanics, that closed subspaces of a Hilbert space $\mathcal{H}$ associated with a quantum system faithfully represent properties of this system, and elements of a unit sphere $S^1(\mathcal{H})$ faithfully represent its pure states, one is led to the following interpretation of values of membership functions of fuzzy sets mentioned in Theorem 1, when it is applied to the family $L(\mathcal{H})$ of closed subspaces of a Hilbert space $\mathcal{H}$ that describes a quantum object:

Let $\mathcal{P}$ be a property of a quantum object, for example, a photon's property of being able to pass through a filter. Let $\hat{P}$ be an operator of orthogonal projection onto a closed linear subspace of $\mathcal{H}$ that, according to the traditional approach, represents the property $\mathcal{P}$.

Then the degree of membership of a state $|\psi\rangle$ to the fuzzy subset P of $S^1(\mathcal{H})$ that collects representations of all states in which the photon is able to pass through the filter equals the probability that a result of an experiment designed to check whether a photon in a state $|\psi\rangle$ possesses the mentioned property, will be positive. Since according to the orthodox quantum theory this probability equals $\langle\psi|\hat{P}|\psi\rangle$, we are bound to accept it as a degree to which the state $|\psi\rangle$ belongs to the subset P:

$$P(|\psi\rangle) = \langle\psi|\hat{P}|\psi\rangle. \tag{5}$$

Let us note that the number (5) represents degree to which a quantum object possesses the property $\mathcal{P}$ *before* an experiment designed to check this property is completed. Since when the experiment is completed, we obtain one of two possible results (e. g., a particular photon either passed or did not pass through the filter) it is tempting to think that a studied quantum object either had or had not (in the "yes–no" sense) the property $\mathcal{P}$ even before the experiment was completed. However, such a position is hard to accept because of already mentioned experimentally confirmed violation of Bell-type inequalities.

When we accept that quantum objects, before experiments designed to check their properties are completed, can possess their properties to the degree that is neither 0 nor 1, we can paraphrase the title of Peres' paper [18] "*Unperformed experiments have no results*" and say: "*Unperformed experiments have all their possible results, each of them to the degree obtained by suitable calculations.*" In particular, a quantum object in a state $|\psi\rangle$ both possess property $\mathcal{P}$ to the degree $P(|\psi\rangle) = \langle\psi|\hat{P}|\psi\rangle$ and does not possess it to the degree $1 - P(|\psi\rangle) = 1 - \langle\psi|\hat{P}|\psi\rangle$.

Similar situations can be found also in the macroworld. Let us consider a biased coin with center of mass so shifted that the probability of landing heads up is 0.7 and the probability of landing tails up is 0.3. Of course, when an experiment (tossing the coin) is completed we obtain one of two possible results. However, no one claims, when the result is, for example, "heads up" that this coin, before the experiment was done, had entirely a property "landing heads up," specially that in the next run of this experiment the result may be different.

Of course when such a "dichotomic" experiment is completed one can unambiguously ascribe all objects that underwent this experiment (mind the past tense!) to one of two crisp sets: coins that landed heads up and coins that landed tails up, according to the results of the experiment. However, *before* the experiment is completed, the degree to which each coin belongs to these sets equals the probability of obtaining respective result. This is an analogy with the famous *wave packet collapse* widely discussed in textbooks on quantum mechanics.

5.3 Interpretation of Łukasiewicz operations

From Theorem 1, it follows that fuzzy set representation of any quantum logic with an ordering set of states, therefore, also an orthomodular lattice $L(\mathcal{H})$ of closed subspaces of a Hilbert space $\mathcal{H}$ that describes properties of a quantum object, is endowed with two pairs of binary operations: order-theoretic operations of *meet* $\wedge$ and *join* $\vee$, and fuzzy set operations of *Łukasiewicz intersection* ⊓ and *Łukasiewicz union* ⊔. Order-theoretic operations are "inherited" from Boolean algebras of crisp subsets of a phase space that describe properties of classical objects, in which case they coincide, respectively, with set-theoretic intersection and union of crisp sets that represent properties of classical objects. For example, if A is a subset of a phase space that collects all states of a classical physical object in which this object has property $\mathcal{A}$, and B is a subset of a phase space that collects all states in which this object has property $\mathcal{B}$, then $A \cap B = A \wedge B$ collects all states in which this object has both properties $\mathcal{A}$ and $\mathcal{B}$, and $A \cup B = A \vee B$ collects all states in which this object has at least one of these properties.

However, in the quantum case when Boolean algebras are replaced by more general orthomodular lattices, this interpretation is doubtful, which was noticed even by the founding fathers of the quantum logic theory. Actually, Birkhoff and von Neumann wrote in [4]:

> *"It is worth remarking that in classical mechanics, one can easily define the meet or join of two experimental propositions as an* experimental proposition *– simply by having independent observers read off the measurements which either proposition involves, and combining the results logically. This is true in quantum mechanics only exceptionally – only when all the measurements involved commute (are compatible)."*

Some light on this problem is shed by theorem proved in [31].

Theorem 2. *Let $\mathcal{L}(\mathcal{U})$ be a quantum logic of fuzzy subsets of a universe $\mathcal{U}$ and let $A, B \in \mathcal{L}(\mathcal{U})$. Then $A \sqcap B \in \mathcal{L}(\mathcal{U})$ iff $A \sqcup B \in \mathcal{L}(\mathcal{U})$, and in this case A and B are compatible, $A \sqcap B = A \wedge B$, $A \sqcup B = A \vee B$.*

Let us note that in any lattice meets and joins are globally defined while, as we already mentioned, in a fuzzy quantum logic that is not a Boolean algebra, Łukasiewicz operations are defined only partially. However, this is rather a virtue than a drawback, since accepting that Łukasiewicz operations, not order-theoretic operations, are basic operations in quantum logics, could help solve the problem that Birkhoff and von Neuman were concerned with.

On the other hand, the fact that order-theoretic operations coincide with Łukasiewicz operations whenever the latter are defined, explains why for so many years order-theoretic operations were regarded as proper tools for constructing mathematical representations of "compund" statements about quantum systems.

To be explicit, consider two properties of a quantum object: property $\mathcal{A}$ and property $\mathcal{B}$. In the traditional model, these properties are represented, respectively, by closed subspaces $\mathbf{A}$ and $\mathbf{B}$ of a Hilbert space $\mathcal{H}$ used to describe this object. Let us now consider "compound" property $\mathcal{C}$ defined as follows:

> "An object has property $\mathcal{C}$ iff it has both properties $\mathcal{A}$ and $\mathcal{B}$."

In the traditional model, the closed subspace $\mathbf{C} \subset \mathcal{H}$ that mathematically represents property $\mathcal{C}$ is equal to the meet of mathematical representations of properties $\mathcal{A}$ and $\mathcal{B}$:

$$\mathbf{C} = \mathbf{A} \wedge \mathbf{B},$$

which is doubtful when orthogonal projections onto these subspaces $\hat{A}$ and $\hat{B}$ do not commute (an old Birkhoff–von Neumann problem). Nevertheless, since the family of all closed subspaces of a Hilbert space $L(\mathcal{H})$ is a lattice, this meet always exists, in spite of the fact that properties $\mathcal{A}$ and $\mathcal{B}$ may be not simultaneously measurable.

In the propounded "fuzzy set" model properties $\mathcal{A}$ and $\mathcal{B}$ are represented, respectively, by fuzzy subsets A and B of a unit sphere $S^1(\mathcal{H})$, and property $\mathcal{C}$ is represented by a fuzzy subset C that is Łukasiewicz intersection of fuzzy subsets A and B:

$$C = A \sqcap B,$$

provided that it belongs to the fuzzy set representation of a lattice $L(\mathcal{H})$. Formally, this Łukasiewicz intersection can be always calculated by formulas (2) and (5), but this does not mean that the fuzzy set C necessarily belongs to the fuzzy set representation of a lattice $L(\mathcal{H})$. In particular, from Theorem 2 it follows that if fuzzy sets A and B are not compatible, that is, operators of orthogonal projection onto subspaces $\mathbf{A}, \mathbf{B} \in L(\mathcal{H})$ do not commute, Łukasiewicz intersection $A \sqcap B$ and union $A \sqcup B$ do not belong to the fuzzy set representation of $L(\mathcal{H})$.

On the other hand, if they do belong, then by Theorem 2 they coincide with meet $A \wedge B$ and join $A \vee B$. Of course, these order-theoretic operations are defined by the partial-order relation that coincides with subset relation on fuzzy sets, but due to the isomorphism proved in Theorem 1, they can be identified with original order-theoretic operations on the orthomodular lattice $L(\mathcal{H})$.

5.4 Other attempts to apply fuzzy set ideas in quantum mechanics

There are already dozens of papers in which adjectives "fuzzy," "vague," or "unsharp" are combined with "quantum" (this can be easily checked by searching in the internet). However, a significant number of these papers are highly speculative, or their relationship to Zadeh's original ideas is rather loose, or their authors are not sufficiently proficient in quantum mechanics.

In this section, we try to indicate papers that are both mathematically and physically sound. This list is certainly far from being exhaustive, and we apologize for any omission.

Eduard Prugovečki was probably the first who applied Zadeh's original ideas in quantum mechanics. His papers [19–24] are based on the idea of a *fuzzy phase space* consisting of Gaussian-shaped *fuzzy points*. However, in his numerous later papers he abandoned this name in favor of a “stochastic phase space.”

Wawrzyniec Guz was inspired by Prugovečki during his staying in Toronto. Since his original field of interest was quantum logic, his two papers, [14] and [15], can be classified as the first papers belonging to the domain of fuzzy quantum logic.

Sławomir Bugajski in a series of papers [5–8] (see also [2, 9]) developed fuzzy probability theory suitable for description of quantum phenomena. In particular, he showed that within this model quantum mechanical observables are in fact fuzzy random variables.

Alex Granik and H. John Caulfield [12] presented the idea that microobjects are “fuzzy” in a sense that they reside simultaneously in various places in the space which, according to the traditional approach, was expressed by saying that there are various probabilities of finding a microobject in various places in the space. In this respect, that is, interpreting quantum-mechanical probabilities as degrees of membership to fuzzy sets, their approach coincides with ours and also with ideas presented in [10].

Although Paul Busch and Gregg Jaeger [10] did not refer at all to the very notion of a fuzzy set, it is clear that their paper could be “translated” into the “proper” language of fuzzy set theory.

The paper [39] by Apostolos Syropoulos is devoted to vagueness in chemistry, but it contains a section interesting to physicists: “Quantum physics and vague chemistry” in which the same idea concerning “fuzzy” interpretation of quantum superpositions as in [35] is expressed.

Out of other papers combining “fuzzy” with “quantum” it is worth to recall papers by Rudolf Seising [38], Mieczyslaw Albert Kaaz [16], Wenbing Qiu [37], Ignazio Licata [17], to mention a few.

It is obvious that a still growing number of papers in which fuzzy set ideas are applied to the explanation of quantum phenomena is a proof that fuzzy sets have found a permanent place in quantum mechanics.

Bibliography

[1] J. S. Bell, On the Einstein Podolsky Rosen paradox, Physics 1 (1964), 195–200.
[2] E. G. Beltrametti and S. Bugajski, The Bell phenomenon in classical frameworks, J. Phys. A, Math. Gen. 29 (1996), 247–261.
[3] E. G. Beltrametti and G. Cassinelli, The Logic of Qantum Mechanics, Addison-Wesley, Reading MA, 1981.
[4] G. Birkhoff and J. von Neumann, The logic of quantum mechanics, Ann. Math. 37 (1936), 823–843.
[5] S. Bugajski, Fuzzy dynamical systems, fuzzy random fields, Rep. Math. Phys. 36 (1995), 263–274.

[6] S. Bugajski, Fundamentals of fuzzy probability theory, Int. J. Theor. Phys. 35 (1996), 2229–2244.
[7] S. Bugajski, Fuzzy dynamics in terms of fuzzy probability theory, in: Proceedings of the Seventh IFSA World Congress, Prague 1997, Vol. IV, 1997, pp. 255–260.
[8] S. Bugajski, Fuzzy stochastic processes, Open Syst. Inf. Dyn. 5 (1998), 169–185.
[9] S. Bugajski, K.-E. Hellwig and W. Stulpe, On fuzzy random variables and statistical maps, Rep. Math. Phys. 41 (1998), 1–11.
[10] P. Busch and G. Jaeger, Unsharp quantum reality, Found. Phys. 40 (2010), 1341–1367.
[11] R. Giles, Łukasiewicz logic and fuzzy set theory, Int. J. Man-Mach. Stud. 67 (1976), 313–327.
[12] A. Granik and H. J. Caulfield, Fuzziness in quantum mechanics, Phys. Essays 9 (2001). arXiv:quant-ph/0107054v1. http://dx.doi.org/10.4006/1.3029260.
[13] D. M. Greenberger, M. Horne and A. Zeilinger, Going beyond Bell's theorem, in: Bell's Theorem, Quantum Theory, and Conceptions of the Universe, M. Kafatos, ed., Kluwer Academic, Dordrecht, 1989, pp. 69–72.
[14] W. Guz, Stochastic phase spaces, fuzzy sets, and statistical metric spaces, Found. Phys. 14 (1984), 821–848.
[15] W. Guz, Fuzzy σ-algebras of physics, Int. J. Theor. Phys. 24 (1985), 481–493.
[16] M. A. Kaaz, Concerning a quantum-like uncertainty relation for pairs of complementary fuzzy sets, J. Math. Anal. Appl. 121 (1987), 273–303.
[17] I. Licata, General systems theory, like-quantum semantics and fuzzy sets, in: Systemics of Emergence. Research and Development, G. Minati, E. Pessa and M. Abram, eds, Springer, Berlin, 2006, pp. 723–734, arXiv:0704.0042.
[18] A. Peres, Unperformed experiments have no results, Am. J. Phys. 46 (1978), 745–747.
[19] E. Prugovečki, Fuzzy sets in the theory of measurement of incompatible observables, Found. Phys. 4 (1974), 9–18.
[20] E. Prugovečki, Measurement in quantum mechanics as a stochastic process on spces of fuzzy events, Found. Phys. 5 (1975), 557–571.
[21] E. Prugovečki, Localizability of relativistic particles in fuzzy phase space, J. Phys. A 9 (1976), 1851–1859.
[22] E. Prugovečki, Probability measures on fuzzy events in phase space, J. Math. Phys. 17 (1976), 517–523.
[23] E. Prugovečki, Quantum two-particle scattering in fuzzy phase space, J. Math. Phys. 17 (1976), 1673–1681.
[24] E. Prugovečki, On fuzzy spin spaces, J. Phys. A 10 (1977), 543–549.
[25] P. Pták and S. Pulmannová, Orthomodular Structures as Quantum Logics, Kluwer, Dordrecht, 1991.
[26] J. Pykacz, Quantum logics as families of fuzzy subsets of the set of physical states, in: Preprints of the Second International Fuzzy Systems Association Congress, Tokyo, July 20–25, 1987, Vol. 2, 1987, pp. 437–440.
[27] J. Pykacz, Fuzzy set ideas in quantum logics, Int. J. Theor. Phys. 31 (1992), 1767–1783.
[28] J. Pykacz, Fuzzy quantum logic I, Int. J. Theor. Phys. 32 (1993), 1691–1708.
[29] J. Pykacz, Fuzzy quantum logics and infinite-valued Łukasiewicz logic, Int. J. Theor. Phys. 33 (1994), 1403–1416.
[30] J. Pykacz, Fuzzy quantum logics as a basis for quantum probability theory, Int. J. Theor. Phys. 37 (1998), 281–290.
[31] J. Pykacz, Łukasiewicz operations in fuzzy set and many-valued representations of quantum logics, Found. Phys. 30 (2000), 1503–1524.
[32] J. Pykacz, "Solution" of the EPR paradox: Negative, or rather fuzzy probabilities?, Found. Phys. 36 (2006), 437–442.

[33] J. Pykacz, Towards many-valued/fuzzy interpretation of quantum mechanics, Int. J. Gen. Syst. 40 (2011), 11–21.
[34] J. Pykacz, Fuzzy sets in foundations of quantum mechanics, in: On Fuzziness: a Homage to Lotfi, A. Zadeh, R. Seising et al., eds, Vol. 2, Springer, Berlin, 2013, pp. 553–557.
[35] J. Pykacz, Quantum Physics, Fuzzy Sets and Logic: Steps Towards a Many-Valued Interpretation of Quantum Mechanics, Springer, Cham, 2015.
[36] J. Pykacz and B. D'Hooghe, Bell-type inequalities in fuzzy probability calculus, Int. J. Uncertain. Fuzziness Knowl.-Based Syst. 9 (2001), 263–275.
[37] W. Qiu, There also can be fuzziness in quantum states itself – breaking through the framework and the principle of quantum mechanics, J. Mod. Phys. 11 (2020), 952–966. http://dx.doi.org/10.4236/jmp.2020.116059.
[38] R. Seising, Can fuzzy sets be useful in the (re) interpretation of uncertainty in quantum Mechanics? in: NAFIPS 2006 Annual Meeting of the North American Fuzzy Information Processing Society, Montreal, 3–6 June 2006, North American Fuzzy Information Processing Society, Montreal, 2006, pp. 414–419, http://dx.doi.org/10.1109/NAFIPS.2006.365445.
[39] A. Syropoulos, On vague chemistry, Found. Chem. 23(1) (2021), 105–113.

Pier Luigi Gentili and Apostolos Syropoulos

6 Vagueness in chemistry

Abstract: Chemistry is intrinsically vague mainly because life is in a way a chemical phenomenon. However, vagueness emerges unexpectedly in other areas of chemistry and this is why molecules can be described by mathematical models of vagueness. Naturally, this means that we must revise the way people understand the discipline and work in it.

6.1 Introduction

Usually, we use the *Aristotelian* or *bivalent* logic, as is also known, to describe natural phenomena. This logic is based on three axioms:
- the law of contradiction, which states $A \wedge \neg A$ is always false;
- the law of identity, which states that A will always be A; and
- the law of excluded middle, which states that $A \vee \neg A$ is always true.

To be fair, we must stress that Aristotle was the first to question the validity of bivalent logic; nevertheless, we do not plan to elaborate on this. It is also true that certain natural phenomena and objects are vague. This idea is known as ontic vagueness. However, this idea is not universally accepted and some scholars and researchers have put forth a number of objections to ontic vagueness (e. g., see [17] for a presentation of such objections). Clearly, if vagueness is something real, then we have to find it or to reveal it. But how can we spot vagueness in a scientific field? Fuzzy set theory is the prevailing theory for the description and analysis of vague phenomena. Therefore, when a fuzzy set is appropriate for describing a facet of a discipline, then that discipline has vague concepts for sure.

The theory of a fuzzy set was introduced by the engineer Lotfi Aliasker Zadeh [21]. Fuzzy sets break the laws of contradiction and excluded middle because an item may belong to both set A and its complement with respect to some universe (i. e., some set X), with the same or different degrees of membership. The degree of membership of an item to a fuzzy set can be any real number of the real interval $[0, 1]$. In chemistry, there are many examples of vague concepts [11, 19]. For example, substances are classified as acids, amphoteres, or bases. However, it is easier to say that there are only acids and bases and all other substances are acids or bases to some degree. Also, in quantum physics we see manifestations of these ideas. For example, the wave-particle dualism and the elementary unit of quantum information (i. e., the qubit) [8, 20]. In what follows, we will give more examples of vagueness in chemistry.

https://doi.org/10.1515/9783110704303-006

Plan of the chapter

In what follows, we examine vagueness at the molecular level and then we discuss vagueness in biochemistry and conclude with a short discussion.

6.2 Molecules as fuzzy sets

If vagueness is everywhere, then it should manifest its presence at the molecular level. Therefore, molecules and their interactions should be vague, nevertheless, it is not clear how these are vague. For example, are water molecules vague? If they are, they should at least have some property to some degree and obviously different molecules of the same type should have this property to a different degree.

The structure of a molecule is known when we define the type and number of atoms that are present, the order and the way the atoms are bound, and the relative disposition of atoms into space. At ordinary temperatures, the available thermal energy allows the same chemical compound to exist as an ensemble of conformers. Conformers are molecules with the same chemical composition and identical bonds but differ in the relative disposition of atoms into the 3D-space. A few examples of conformers of a merocyanine are shown in Figure 1.

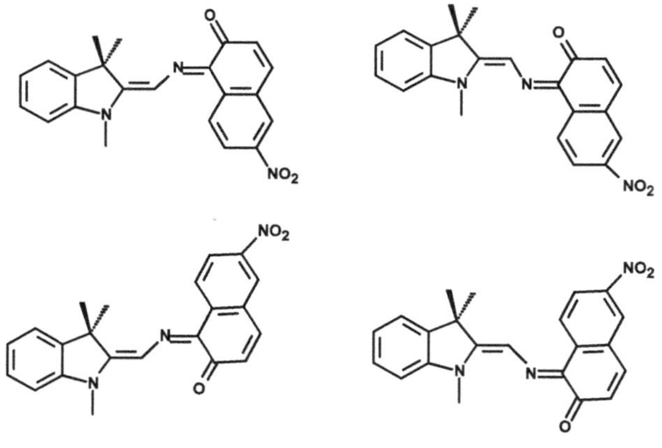

Figure 1: Examples of conformers of a merocyanine.

The types and the relative amounts of the different conformers depend on the physical and chemical contexts. Physicochemical parameters such as temperature, type of solvent, ionic strength, pH, etc., affect the type and number of different possible conformers that are accessible. Conformational heterogeneity and dynamism enable context-specific functions to emerge in response to changing environmental conditions. The same compound can show different behavior depending on the context. From this point of view, each compound is like a word of the natural language, whose meaning is context-dependent. Therefore, an ensemble of conformers for a particular chemical compound works as if they were fuzzy sets [8, 9]. The fuzziness of a macromolecule is usually more pronounced than that of a simpler molecule because it exists in a larger number of conformers. Among proteins, those that are entirely or partially disordered are the fuzziest [6]. Their remarkable fuzziness makes them multifunctional and suitable to moonlight, that is, they play distinct roles, depending on their context [16].

When compounds that exist as a collection of conformers and that respond to the same type of either physical or chemical stimulus are combined, they granulate the variable representing the stimulus in a group of molecular fuzzy sets. Such molecular fuzzy sets are activated at different degrees by distinct stimuli. This strategy is at the core of the sensory subsystems of the human nervous system [10]. For instance, the three types of cones that we have in our retina, that is, the blue, green, and red cones, respectively, partition the visible region into three fuzzy sets, which represent their three absorption spectra, as shown in Figure 2. Stimuli of visible light with different spectral compositions belong to the three molecular fuzzy sets at different degrees. They are perceived as distinct colors by our brains.

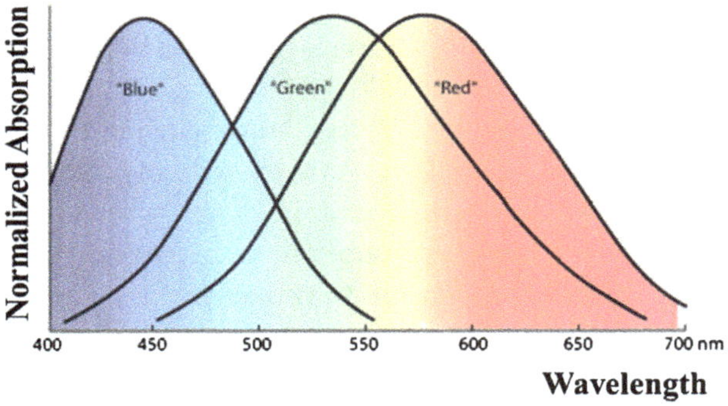

Figure 2: The absorption spectra of the three cones we have on our retina: the blue, the green, and the red cones.

It is worth mimicking the functional way our sensory subsystems work. Its imitation allows the design of artificial sensory systems that are strongly sensitive. For instance, the absorption spectra of the three cones in Figure 2 can be substituted by absorption spectra shifted to the UV and belonging to three direct photochromic compounds. A direct photochromic compound absorbs in the UV and when it is irradiated by UV frequencies, it generates an absorption band into the visible region, giving rise to a specific color. If three distinct photochromes, having different absorption spectra in the UV, are mixed, they partition the UV spectral region in three partially overlapped fuzzy sets. When they are irradiated by a specific UV wavelength, they are activated at different degrees and they originate different absorption spectra in the visible region. Such biologically inspired photochromic fuzzy logic system allows us to extend human color vision from the visible to the UV [12, 13].

6.3 Vagueness in biochemical life

In science, there are foundational concepts such as time, energy, life [7], and biological species, which lack rigorous definitions. A definition of life is required by three categories of scientists:
(I) astrobiologists searching for the presence of extraterrestrial living systems;
(II) scientists involved in research line of artificial life;
(III) prebiotic chemists interested in the origin of life on Earth [3].

The reasonable lack of a definition of life introduces vagueness in the field of astrobiology, artificial life, and prebiotic chemistry. Prebiotic chemistry has been defined as a vague discipline [4], but also astrobiology, and artificial life are vague disciplines. In fact, life can be defined as a multidimensional fuzzy set. Its dimensions coincide with all those features that are shared by the mesmerizing varieties of life forms. These are:
(A) The life cycle: birth; growth; aging; death. It might range from 1 day up to 100 years.
(B) The chemical composition: the elementary unit of every living being is the cell that contains DNA, RNA, proteins, phospholipids, water, salts, ATP, and others.
(C) The shape: it ranges from the shape and micrometric size of a single cell to the bodies of multicellular organisms, which can be as large as tens of meters.
(D) The information: living beings use information for reaching their goals, according to the theory of teleonomy [15]. The basic goals common to all living beings are to survive and reproduce.
(E) Adaptation, acclimation, evolution: every living being is able to adapt, acclimate, and evolve.

If we look at the case studies of prebiotics, astrobiology, and artificial life, we must admit degrees of lifeness (or “livingness”). Viruses and robots belong to the life-fuzzy set with degrees of membership less than one. Prebiotic chemists investigate the phase transition from inanimate to animate matter. This phase transition can be described as a progressive increase of the degree of membership of the self-organizing inanimate matter to the life-fuzzy set. Different degrees of membership of the prebiotic systems to the life-fuzzy set do not reflect a temporal evolution sequence. It is highly probable that the first living cells result from the fortuitous association of systems which, after association could have lost some of their previous characteristics or functions. The idea that systems with a higher degree of membership necessarily appeared, during prebiotic evolution, after systems characterized by lower degrees cannot be considered as a general rule [4]. The transition from nonliving to living might have occurred progressively through systems which were "not yet living" but already "not fully nonliving." It is generally accepted that the emergence of chemical systems exhibiting basic life-like properties, including self-maintained chemical networks, self-replicative polymers, and self-reproductive vesicles appear to have been necessary steps during the prebiotic phase [18]. Eventually, these three prebiotic systems somehow articulated harmoniously and the first living beings popped up.

6.4 Discussion

Fuzzy chemistry [5] was an attempt to develop a formal theory of general chemistry. However, the theory describes the real world only if its mathematical structure is based on fuzzy arithmetics. This simply means that vagueness is something fundamental. This is the reason why chemistry, which is considered to be an exact science, is rarely an *always* or *never* science [2]. The fact that many considered fuzzy mathematics as the ideal tool to work in chemistry (e. g., see [19]) is yet another "proof" that chemistry is intrinsically vague. And of course one should not forget that the Hughes–Ingold theory of nucleophilic substitution in organic chemistry [14] can be best described using fuzzy mathematics [1]. These facts suggest that chemistry is a discipline that is vague so people should adopt the way they tackle and solve problems in chemistry.

Bibliography

[1] F. M. Akeroyd, Why was a Fuzzy Model so Successful in Physical Organic Chemistry? HYLE: Int. J. Philos. Chem. 6(2) (2000), 161–173. Paper available online from http://www.hyle.org/journal/issues/6/akeroyd.htm.

[2] J. Bernstein, Structural Chemistry, Fuzzy Logic, and the Law, Isr. J. Chem. 57(1–2) (2017), 124–136.

[3] G. Bruylants, K. Bartik and J. Reisse, Is it Useful to Have a Clear-cut Definition of Life? On the Use of Fuzzy Logic in Prebiotic Chemistry, Orig. Life Evol. Biosph. 40(2) (2010), 137–143.

[4] G. Bruylants, K. Bartik and J. Reisse, Prebiotic chemistry: A fuzzy field, C. R., Chim. 14(4) (2011), 388–391, De la chimie de synthèse à la biologie de synthèse.

[5] G. Cerofolini and P. Amato, Fuzzy Chemistry: An Axiomatic Theory for General Chemistry, in: IEEE International Fuzzy Systems Conference, 2007 IEEE International Fuzzy Systems Conference, IEEE, London, 2007, pp. 1–6.

[6] M. Fuxreiter, Towards a Stochastic Paradigm: From Fuzzy Ensembles to Cellular Functions, Molecules 23 (2018), 11, Article number: 3008.

[7] J. Gayon, C. Malaterre, M. Morange, F. Raulin-Cerceau and S. Tirard, Defining Life: Conference Proceedings, Orig. Life Evol. Biosph. 40(2) (2010), 119–120.

[8] P. L. Gentili, The Fuzziness of the Molecular World and Its Perspectives, Molecules 23(8) (2018), 2074.

[9] P. L. Gentili, The fuzziness of a chromogenic spirooxazine, Dyes Pigments 110 (2014), 235–248. 1st International Caparica Conference on Chromogenic and Emissive Materials.
[10] P. L. Gentili, The human sensory system as a collection of specialized fuzzifiers: A conceptual framework to inspire new artificial intelligent systems computing with words, J. Intell. Fuzzy Syst. 27(5) (2014), 2137–2151.
[11] P. L. Gentili, The Fuzziness in Molecular, Supramolecular, and Systems Chemistry, MDPI, Basel, 2020.
[12] P. L. Gentili, A. L. Rightler, B. M. Heron and C. D. Gabbutt, Discriminating between the UV-A, UV-B and UV-C regions by novel Biologically Inspired Photochromic Fuzzy Logic (BIPFUL) systems: A detailed comparative study, Dyes Pigments 135 (2016), 169–176, Special Issue: 2nd International Caparica Conference on Chromogenic and Emissive Materials 2016.
[13] P. L. Gentili, A. L. Rightler, B. M. Heron and C. D. Gabbutt, Extending human perception of electromagnetic radiation to the UV region through biologically inspired photochromic fuzzy logic (BIPFUL) systems, Chem. Commun. 52(7) (2016), 1474–1477.
[14] C. K. Ingold, Principles of an Electronic Theory of Organic Reactions, Chem. Rev. 15(2) (1934), 225–274.
[15] J. Monod, Chance and necessity: An essay on the natural philosophy of modern biology, Knopf, New York, 1971.
[16] C. Jeffery, An introduction to protein moonlighting, Biochem. Soc. Trans. 42(6) (2014), 1679–1683.
[17] G. Keil, Introduction: Vagueness and Ontology, Metaphysica 14(2) (2013), 149–164.
[18] J. Peretó, Out of fuzzy chemistry: from prebiotic chemistry to metabolic networks, Chem. Soc. Rev. 41(16) (2012), 5394–5403.
[19] D. H. Rouvray, Fuzzy Logic in Chemistry, Academic Press, San Diego, 1997.
[20] A. Syropoulos, On vague chemistry, Found. Chem. 21(1) (2021), 105–113.
[21] L. A. Zadeh, Fuzzy Sets, Inf. Control 8 (1965), 338–353.

Bjørn Hofmann

7 Vagueness in medicine: on indistinct phenomena, vague concepts, and uncertain knowledge

Abstract: In this chapter, I will examine vagueness in medicine in five manners. First, medicine is a discipline with unclear borders, as it builds on a wide range of other disciplines and subjects. Second, medicine deals with a wide range of indistinct phenomena resulting in borderline cases. Third, medicine uses many vague concepts, making it unclear which situations, conditions, and processes that fall under them. Fourth, medicine is based on and produces uncertain knowledge and evidence. Fifth, vagueness emerges in medicine as a result of a wide range of fact-value-interactions. Accordingly, there are many kinds of vagueness in medicine: disciplinary, ontological, conceptual, epistemic, and fact-value-related vagueness.

7.1 Medicine as a vague discipline

Initially, it is important to acknowledge that medicine is not a well-defined discipline. It builds on a great variety of other subjects, such as biology, chemistry, physics, and physiology. It is not always clear what belongs to medicine and what belongs to other disciplines. This generates a lot of borderline cases. Additionally, medicine is considered to be both a theory and a practice, and as a practice it is both clinical and experimental. Furthermore, medicine has many goals, and the borders between them are not clear. Medicine aims at curing, preempting, predicting, palliating, and understanding disease, as well as caring for the well-being of the person [1].

Hence, medicine is in itself a vague discipline. Moreover, what we mean by vagueness may itself be vague [2]. That is, "vagueness" in "vagueness in medicine" is vague. While both these issues are interesting and have been studied elsewhere, I will leave them for later studies and focus on the various types of vagueness at play in the activity which falls under ordinary use of the concept of medicine.

7.2 Ont(olog)ical vagueness: continuity

Vagueness is characterized by borderline cases [3]. This is highly evident in medicine, dealing with individual persons with unique life stories and exclusive biological, men-

Note: This chapter is based on a submitted and revised article manuscript with the title "Vagueness in Medicine: On disciplinary indistinctness, fuzzy phenomena, vague concepts, uncertain knowledge, and fact-value-interaction"

https://doi.org/10.1515/9783110704303-007

tal, and social makeup. Even well-described and well-defined diseases appear differently in individuals.

Moreover, the basic phenomena of medicine, such as pain, suffering, harm, but also wellness and well-being are vague phenomena. For example, pain comes in many kinds and grades and is hard to assess and measure. The same goes for suffering [4] and well-being. Moreover, according to Jennifer Worrall and John Worrall the nature of disease is unclear: "there is no such thing as disease. The notion fails to find any joints in nature at which to carve" [5]. According to this way of thought, there is no disease in nature, only human significance attached to certain conditions [6].

The challenge of continuity is visible with respect to other (and more technical) phenomena in medicine, such as dysfunction, symptoms, and signs. What counts as a sign of disease in one context, such as an enlarged heart, can be a sign of wellness and high performance in another, such as in an athlete. Even more, it is argued that "many medical objects and subjects, for example, cells, tissues, organs, organisms, persons, patients, symptoms, diseases, individual disease states of patients, and recovery processes are ontically vague to the effect that their vagueness is principally not eliminable" [7].

Hence, many of the basic phenomena (that are investigated and manipulated) in medicine are vague. So are the phenomena that comprise medicine's goal. Therefore, there is a basic vagueness inherent in the theory and practice of medicine.

7.3 Conceptual vagueness

In addition to (and partly related to) the ontological vagueness there is a conceptual vagueness in medicine. Basic concepts, such as health and disease do not have clear boundaries, are based on other vague concepts (such as wellbeing and happiness), and are hard to define [8–24]. There are ample borderline cases [25].

Accordingly, it has been maintained widely that disease is a vague concept [26–28]. "The meaning of the word 'disease' is vague, elusive, and unstable. You may reach for it, but you won't grasp it, except in pieces and fragments. It is prone to changes, permutations, and shatterings – according to the circumstances, or irrespective of them" [29]. The same claim of vagueness goes for other health related concepts, such as sickness and illness [30].

In general, a concept is considered to be vague if no additional criteria will improve its definition and sharpen its boundaries [31]. Vague concepts have borderline cases that resist investigation [3]. In a *relative borderline case*, the concept (of disease) is clear, but the means to decide whether or not a condition falls in under the concept (i. e., whether it is disease) are incomplete. In an *absolute borderline case*, no amount of conceptual analysis or empirical studies can settle the issue. Accordingly, conceptual vagueness can be due to confused perception of a real concept or because of vagueness in its extension: "first... that 'disease' is indeed a real concept, a 'natural kind,' but the perception of this real concept even amongst experts is vague and confused. ... [or] second ... that there is no real distinction in nature between 'disease' and 'non-disease' " [5].

In medicine, a wide range of controversial cases have been discussed, such as baldness, skin wrinkles, cellulitis, freckles, jet lag, ear wax accumulation, teeth grinding, chronic fatigue syndrome, and fibromyalgia [32]. More recently, gender identity disorder and gender incongruence have been heatedly debated [33–37].

However, it is not only the basic concepts of medicine, such as health and disease that are vague. Many terms in medicine are vague [7]. Deciding what falls under concepts like pneumonia, Alzheimer's disease, and schizophrenia is not easy as there are ample borderline cases. Moreover, medical descriptions include vague predicates (pain, headache, icterus), quantifies (few, many, most), temporal notions (acute, chronic, rapid), and vague frequency notions (commonly, usually, often) [7].

Take cancer as an example. It is defined by the National Cancer Institute as "[a] term for diseases in which abnormal cells divide without control and can invade nearby tissues" (https://www.cancer.gov/). However, for many cancers there are different stages that more or less satisfy the definition. Ductal carcinoma in situ (DCIS) is but one example. While the cells in DCIS may divide without control, they do not necessarily invade nearby tissue [38].

Because key concepts in medicine have practical implications, it is crucial to clarify their boundaries. For example, whether your condition falls under the concept of disease decides whether you get attention and care from health providers, whether you are freed from social obligations (to work), and whether you will get economic support (sickness benefits). Setting such limits have been called the line-drawing problem and is widely discussed in the philosophy of medicine [39, 40]. Hence, conceptual vagueness is a key challenge in medicine.

7.3.1 Different concepts of malady

One source of elusiveness in medical semantics is the imprecision and ambiguity of terms. Disease, illness, sickness, impairment, disorder, and dysfunction are frequently used interchangeably. However, some clarity has been provided in differentiating between illness, disease, and sickness. *Illness* has been defined as the personal experience of a (mental or bodily) condition that is considered to be harmful while *disease* are conditions that health professionals consider to be harmful or disadvantageous to a person (called patient), and *sickness* is the social role attributed to a person assumed to have a disease or being ill [41–43]. All three fall under the concept of malady [44].

The relationship between the three basic perspectives on human malady is illustrated in Figure 1. As illustrated in the figure, the terms are interrelated and partly interdependent [42]. However, they only partly overlap, resulting in both in disagreement between the three perspectives on human malady and generating borderline cases.

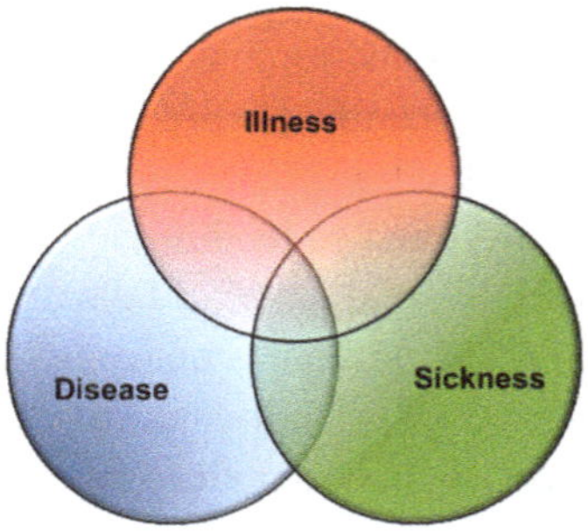

Figure 1: Three perspectives on human malady. Adapted from [42].

The distinctions of the concepts of disease, illness, and sickness can explain many misunderstandings between health professionals and patients. For example, the physician can say that "I cannot find disease" (meaning "I cannot identify any known diseases to the condition that you have") while the patient hears "you are not ill" (meaning that "I do not believe that you experience illness").

7.3.2 Polyvalent concepts in medicine

While the distinction between disease, illness, and sickness is helpful in avoiding confusion, the terms are not always used consistently. For example, in psychiatry diseases are frequently called "disorders" and are classified as "illnesses" [10, 45–51]. Hence, there is still ample bewilderment.

Another source of confusion is the fact that health related terms, such as disease, are used in many ways. When a person says that "I have a disease" (when talking about his coughing), he uses the term "disease" to explain why he is coughing. However, when the person maintains that "chronic fatigue syndrome is a disease," he uses it to claim that the condition described as CFS gives certain rights. Correspondingly, when a person says "I feel sick" and "we live in a sick society," the person means different things with "sick."

Another reason why the concept of disease has been conceived of as "vague" is that it can be understood in different ways, for example, depending on knowledge and competence of the concept possessor [52]. The physician and the patient may have different understandings of "arthritis", but they may both have the same concept [53].

The physician knows much more about the pathophysiological conditions related to "arthritis," and the patient more about the first-person experience of "arthritis." Accordingly, the vagueness is claimed to appear because people understand basic concepts, such as "disease," differently. In her study of atherosclerosis, Mol (2002) found that the disease appears as many things in many contexts and refers to "a co-existence of multiple entities with the same name" [54]. Accordingly, there are many borderline issues between the different understandings of such basic concepts in medicine.

Another source of nonclarity is the lack of conceptual unity. As argued by Robert Evan Kendell: "Most physicians ... use the words disease and illness in different senses at different times" [55]. Correspondingly, Carl Jaspers has claimed that disease is not unitary concept, but that there exist many concepts of disease, which in principle can be sharp, but are not in practice [56]. While some philosophers of medicine have been concerned with one conception of disease, for example the pathological concept of disease [57], they have willfully ignored others, such as clinical, judicial, and insurance-related notions [5]. Others have conceived of concepts such as health and disease "as a 'family' of concepts" [58] or as models for searching for "unnoticed causal factors and expressions of disease" [59]. The point is that these conceptions are not unitary and leave lots of borderline cases.

Another level of confusion stems from the attempt to address different value settings [60], perspectives [61], and theoretical aspirations [62]. Accordingly, it is argued that "at the present time, there is no unified concept of disease... In diagnosing different diseases, we often use entirely different types of fundamental criteria. We may have a concept of disease ... based upon gross anatomical defects, microscopic anatomical changes, so-called specific etiological agents, specific deficiencies, genetic aberrations, physiological or biochemical abnormalities, constellations of clinical symptoms and signs, organ and system involvements, and even just descriptions of abnormalities" [63].

In sum, it can be argued with Herbert George Wells that most terms are vague: "... every term goes cloudy at its edges... Every species waggles about in its definition, every tool is a little loose in its handle, every scale has its individual" [64].

While precisification can reduce or remove ambiguity and explanation can clarify complexity, conceptual vagueness may still leave many borderline cases [3], which can confuse the communication between health professionals and patients [65].[1] Attempts have been made to address such challenges, for example, by defining disease as a concept of fuzzy disease [66] based on the prototype resemblance theory of disease [67]. However, a lot of vagueness still prevails.

1 I have not differentiated between linguistic vagueness (such as vague terms) and semantic vagueness (vagueness in representation). See for example reference [7].

7.3.3 Expanding the borders of medicine: expansion of the concept of disease

As vagueness is defined in terms of borderline cases and one of the crucial and delimiting concepts in medicine is disease, it is especially interesting to look at the border of this concept. Moreover, it is widely argued that this border has been expanded significantly [12, 18, 68–72]. Table 1 summarizes six ways that disease has expanded and examples of how this extends its vagueness.

Table 1: Six types of expansion of disease with descriptions, examples, and kinds of vagueness. Adapted and expanded from [73].

Type of expansion	Description	Example	Kind of vagueness
Medicalization	Including ordinary life experiences [74]	Grief, sexual orientation (homosexuality)	Ontological vagueness
Overdiagnosis	Labeling indolent conditions as disease [75]	Ductal carcinoma in situ (DCIS)	Conceptual and epistemic vagueness
Aesthetic expansion	Treating aesthetic characteristics as disease [76]	Protruding ears	Conceptual vagueness
Pragmatic expansion	Making something disease because it can be detected and treated [77]	Hypertension, hyperglycemia	Epistemic vagueness
Conceptual expansion	Expanding definitions of disease [78]	Prediabetes, pre-Alzheimer, making menopause or aging a disease	Conceptual vagueness
Ethical expansion	Making something disease because that will provide attention and access to care	Obesity [79], Attention Deficit Hyperactivity Disorder (ADHD), gender incongruence [80]	Epistemic vagueness Fact-value-vagueness

This chapter only leaves room for mentioning one type of expansion in some detail. The concept of disease was originally intrinsically defined by manifest conditions (such as visible infections and palpable tumors) and related, observed, and/or expressed suffering. In addition, diseases were given names that were called "diagnoses."

However, with time (and expanded knowledge) many conditions where defined as disease without manifest conditions or observed suffering. For example, precursors, risk factors, predictors, and indicators were included to fall under the term disease. Moreover, conditions were called disease beyond what was experienced as disadvantageous or harmful [81, 82].

Having a diagnosis was extended beyond having a disease, and having a disease was expanded beyond what was observed or experienced. This has vastly extended the borderline cases of medicine, and hence, its vagueness.

7.4 Epistemic vagueness: uncertainty

As stated in the quote attributed to Voltaire (François-Marie Arouet), medicine has been (and still is) haunted by uncertainty: "Doctors are men who prescribe medicines of which they know little, to cure diseases of which they know less in human beings of whom they know nothing" [83]. Uncertainty in medicine partly results from the types of vagueness discussed above, but also from ample borderline cases in medical knowledge production.

As with science-based decision-making in general [84], there are many types of uncertainty at play in medicine [85]. According to Sven Ove Hansson, there is: (1) Uncertainty about which alternatives that exist, (2) Uncertainty about the consequences of alternatives, (3) Uncertainty about whether information is trustworthy, and (4) Uncertainty about values and conceptions among decision makers [86]. Moreover, Hansson identifies a continuity from disease-centered to person-centered uncertainty, with scientific (data-centered) uncertainty on the one side, practical (system-centered uncertainty) in the center, and personal (patient-centered) uncertainty on the other [87].

There are many ways to understand uncertainty in medicine. One fruitful framework differentiates between risk, fundamental uncertainty, ignorance, and indeterminacy (Van Asselt 2000; Wynne 1992). *Risk* is when you know certain outcomes and the chance that they occur. Given specific findings for a diagnostic test and the test characteristics (sensitivity, specificity), you know the risk that the patient has a given disease. Correspondingly, the chance of specific benefits and harms of various examinations and treatments is known, and hence the corresponding risk. The same goes for the chance of certain diseases to develop in specific manners (prognosis) and the chance of getting certain diseases for various groups in the population (epidemiology).

According to *fundamental uncertainty* (also known as severe or Knightian uncertainty) you still know about the outcome (e. g., a given disease) but you do not know the probability (distribution). For example, the chance of specific outcomes of diagnostics and treatments may be uncertain. While you have certain signs and symptoms, you do not always know whether they will develop to manifest disease and suffering (progression uncertainty). Moreover, you may also not know the chances that specific indicators, such as precursors, predictors, or risk factors, will develop into manifest disease (development uncertainty). This results in both underdiagnosis and overdiagnosis. Additionally, diagnostic tests frequently result in incidental findings of unknown implications.

Ignorance are unknown factors that are relevant for the diagnostic, prognostic, or therapeutic process, but which the health professional is not aware of. Before Wilhelm Conrad Röntgen's discovery in 1895 people were ignorant of X-rays as medical doctors were of the side-effects of the drug thalidomide. Ignorance can be due to unknown effects of treatments (good or bad), unknown meaning of certain markers for diagnosis or prognosis, and due to the unknown relevance of individual health data.

Indeterminacy, which formally is a type of model validity uncertainty, is uncertainty stemming from different ways to classify and categorize the medical phenomena. As discussed above, defining disease(s), signs, and symptoms is not easy, and many (if not most) of the entities, phenomena, measures in medicine are vague. Defining and measuring personal experience (illness) is challenging. Moreover, there is vagueness in description of findings, in reporting diagnostic tests, and in interpreting and reporting treatment results. Correspondingly, it is difficult to define what is clinically relevant.

Table 2 gives an overview of these four types of uncertainty with examples.

The types of uncertainty mentioned here have many sources and implications. They can be recognized as and resulting in vagueness in the production of medical knowledge and evidence. Some argue that medical nihilism is the most relevant response to vagueness in medical evidence production [90] while others confer to medical cosmopolitanism [1]. Reactions to vagueness is beyond the scope of this chapter.

Moreover, while some count epistemic vagueness as vagueness [91] others are more hesitant [3]. In this section, I have taken as a point of departure that the production of medical knowledge and evidence involves and leaves a wide range of borderline cases.

7.5 Vagueness due to fact-value interactions

The last type of vagueness to be discussed in this chapter is a vagueness resulting from fact-value relationships. Many facts in medicine are value-related leaving ample of borderline issues. In diagnostics, disease definitions, and treatment outcome measures there are thresholds and cut-offs that are not given by nature but decided upon by professionals (and others) from what is believed to give most benefit (or profit).

For example, the various types of uncertainty discussed above result in tradeoffs between Type I and Type II errors. In particular, the cut-off values in diagnostic tests is a tradeoff between test sensitivity and specificity and depend on how you value false positive test results compared to a false negative ones.

Correspondingly, between normal and pathological there are great many borderline cases. Drawing the line for what is normal is not only a factual issue but involves

Table 2: Four types of uncertainty classified according to outcomes and risks. Adapted from [88] and [89].

Possibilities probability	**Known outcome**	**Unknown outcome**
Known probability	Risk	Indeterminacy (Ambiguity, Vagueness)
	Test accuracy (sensitivity, specificity, predictive values) for the various examinations in different contexts.	Defining disease(s), signs, and symptoms
	Chance of specific benefits and harms of various examinations and treatments	How to define and classify specific entities, phenomena, measures
	Chance of disease development (prognosis)	Defining and measuring personal experience (illness)
	Chance of getting certain diseases	Vagueness in description of findings or in reporting of treatment results
		Defining clinical relevance
Unknown probability	Fundamental (Knightian) Uncertainty	Ignorance
	Unknown chance of specific outcomes of diagnostics and treatments Prognostic uncertainty	Unknown effects of treatments (good or bad)
	– Development uncertainty – Progression uncertainty	Unknown meaning of certain markers for diagnosis or prognosis
	Overdiagnosis, underdiagnosis	Unknown relevance of individual health data
	Incidental findings of unknown implications	

conceptions of good and bad. While this is most obvious in psychiatry [92], the same goes for somatic medicine. Correspondingly, and as already alluded to, there are also a wide range of borderline cases between health and disease [93]. People consider themselves healthy although they have many diagnoses and are on medications for several diseases.

One source of vagueness is empirical underdetermination and what has been called "inductive risk" and "epistemic risk." Originally "inductive risk" was identified in the decision of how much evidence is enough to accept or reject a hypothesis [94], but has been extended to include the choice of methodology (e. g., the setting of a level of statistical significance), the characterization of evidence, and the interpretation of data [95], as well as the choice of model organism [96]. See also [97–99]. The point is that all the choices that are involved in the production of evidence in medicine, are value related. This can be seen in the field of health technology assessment where a wide range of (moral) value-judgments have been identified in every step of the production and assessment of evidence [100, 101].

The vagueness due to the fact-value interrelatedness partly stems from many medical concepts being what Bernard Williams called "thick concepts" [102]. According to Williams, thick concepts express a union of fact and value and have extensions which are codetermined by both their descriptive and their evaluative meaning. "Any such concept, . . . , can be analyzed into a descriptive and a prescriptive element: it is guided round the world by its descriptive content, but has a prescriptive flag attached to it. It is the first feature that allows it to be world-guided, while the second makes it action-guiding" [103]. Following Williams, you cannot decide on the extension of "disease" on basis of only its descriptive content. The combination of descriptive and normative content make some characterize it as vague [104].

7.6 Vagueness in diagnosis

The very last area of vagueness in medicine to be mentioned and which stems from the fact-value interaction is vagueness in diagnosis. Classifications of diseases (taxonomies, nosologies) are vague in the sense that they leave many borderline cases. A patient may not fully qualify for a specific diagnosis or be on the borderline of two or more diagnoses.

On the one hand, the diagnostic criteria may be vague making it unclear whether a person falls under the diagnosis or not. On the other hand, the criteria may be clear, but it is unclear whether it applies to a specific person (e. g., due to uncertainty).

While fuzzy reasoning has been suggested to address the vagueness in medical diagnosis [67] others argue we should accept and even praise vagueness [105, 106]. Again, how to handle vagueness in medicine is beyond the scope of this chapter. Another issue that could not find place is the relationship between vagueness, paradoxes, antinomies, and aporias [107].

7.7 Conclusion

In this chapter, I have discussed vagueness in medicine in five manners. First, medicine is a discipline with unclear borders, as it builds on a wide range of other disciplines and subjects. Second, medicine deals with a wide range of indistinct phenomena resulting in borderline cases. Third, medicine uses many vague concepts, making it unclear which situations, conditions, and processes that fall under them. Fourth, medicine is based on and produces uncertain knowledge and evidence. Fifth, vagueness emerges in medicine as a result of a wide range of fact-value interactions. Hence, there are many variants of vagueness present in medicine as elsewhere in life: disciplinary, ontological, conceptual, epistemic, and fact-value related vagueness.

Bibliography

[1] A. Broadbent, Philosophy of medicine, Oxford University Press, 2019.

[2] I. Hu, 'Vague' at Higher Orders, Mind 126(504) (2017), 1189–1216.

[3] R. Sorensen, Vagueness and contradiction, Clarendon Press, 2001.

[4] B. Hofmann, Suffering: Harm to Bodies, Minds, and Persons, in: Handbook of the Philosophy of Medicine, T. Schramme and S. Edwards, eds, Springer Netherlands, Dordrecht, 2017, pp. 129–145.

[5] J. Worall and J. Worall, Defining disease: much ado about nothing?, Analecta Husserl. 72 (2001), 33–55.

[6] P. Sedgwick, in: Illness –- mental and otherwise, The Hastings Studies Center Studies, Vol. 3, 1973, pp. 19–40.

[7] K. Sadegh-Zadeh, Handbook of Analytic Philosophy of Medicine, Vol. 119, Springer, Dordrecht, 2015.

[8] J. Bircher, Towards a dynamic definition of health and disease, Med. Health Care Philos. 8(3) (2005), 335–341.

[9] B. Brülde, On how to define the concept of health: A loose comparative approach, Med. Health Care Philos. 3(3) (2000), 303–306.

[10] B. Brülde, On defining "mental disorder": Purposes and conditions of adequacy, Theor. Med. Bioethics 31(1) (2010), 19–33.

[11] E. J. Campbell, J. G. Scadding and R. S. Roberts, The concept of disease, Br. Med. J. 2(6193) (1979), 757–762.

[12] J. Doust, J. M. Walker and W. A. Rogers, Current Dilemmas in Defining the Boundaries of Disease, J. Med. Philos. 42(4) (2017), 350–366.

[13] M. Ereshefsky, Defining 'health' and 'disease', Stud. Hist. Philos. Biol. Biomed. Sci. 40(3) (2009), 221–227.

[14] M. B. First and J. C. Wakefield, Defining 'mental disorder' in DSM-V. A commentary on: 'What is a mental/psychiatric disorder? From DSM-IV to DSM-V' by Stein et al. (2010), Psychol. Med. 40(11) (2010), 1779–1782, discussion 1931-4.

[15] B. Hofmann, The concept of disease – vague, complex, or just indefinable?, Med. Health Care Philos. 13(1) (2010), 3–10.

[16] J. Kovacs, The concept of health and disease, Med. Health Care Philos. 1(1) (1998), 31–39.

[17] H. Merskey, Variable meanings for the definition of disease, J. Med. Philos. 11(3) (1986), 215–232.

[18] R. Moynihan, Medicalization. A new deal on disease definition, BMJ 342 (2011), d2548.

[19] L. T. Niebroj, Defining health/illness, social and/or clinical medicine?, Silesian Medical University, Poland, J. Philos. Pedagogy 57(4) (2006).

[20] T. Parsons, Definitions of Health and Illness in the Light of American Values and Social Structure, in: Patients, physicians and illness: sourcebook in behavioral science and medicine, E. G. Jaco, ed., Free Press, New York, 1958, pp. 165–187.

[21] P. H. Schwartz, Decision and Discovery in Defining 'Disease', in: Establishing Medical Reality: Essays In The Metaphysics And Epistemology Of Biomedical Science, H. Kincaid and J. McKitrick, eds, Springer Netherlands, Dordrecht, 2007, pp. 47–63.

[22] P. H. Schwartz, Progress in Defining Disease: Improved Approaches and Increased Impact, J. Med. Philos. 42(4) (2017), 485–502.

[23] L. K. F. Temple, R. S. McLeod, S. Gallinger and J. G. Wright, Defining disease in the genomics era, Science 293(5531) (2001), 807–808.

[24] J. Worrall and J. Worrall, Defining Disease: Much Ado about Nothing? in: Life Interpretation and the Sense of Illness within the Human Condition, Analecta Husserliana, Vol. 72, A.-T. Tymieniecka and E. Agazzi, eds, Springer, Netherlands, 2001, pp. 33–55.

[25] B. Hofmann, Simplified models of the relationship between health and disease, Theor. Med. Bioethics 26(5) (2005), 355–377.
[26] R. Rosenberg, Some themes from the philosophy of psychiatry: a short review, Acta Psychiatr. Scand. 84(5) (1991), 408–412.
[27] K. Sadegh-Zadeh, Toward metamedicine, Edit. Metamed. 1 (1980), 3–10.
[28] H. Nordby, The importance of knowing how to talk about illness without applying the concept of illness, Nurs. Philos. 5 (2004), 30–40.
[29] P. Sundström, Disease: The phenomenological and conceptual center of practical-clinical medicine, in: Handbook of phenomenology and medicine, Springer, 2001, pp. 109–126.
[30] K. W. M. Fulford, Moral Theory and Medical Practice, Cambridge University Press, Cambridge, 1989.
[31] T. Williamson, Vagueness, Routledge, London, 1994.
[32] R. Smith, In search for "non-disease", BMJ 324 (2002), 883–885.
[33] M. E. Castro-Peraza, J. M. García-Acosta, N. Delgado, A. M. Perdomo-Hernández, M. I. Sosa-Alvarez, R. Llabrés-Solé and N. D. Lorenzo-Rocha, Gender identity: the human right of depathologization, Int. J. Environ. Res. Public Health 16(6) (2019), 978.
[34] J. Drescher, Controversies in gender diagnoses, LGBT Health 1(1) (2014), 10–14.
[35] T. F. Beek, P. T. Cohen-Kettenis and B. P. C. Kreukels, Gender incongruence/gender dysphoria and its classification history, Int. Rev. Psychiatry 28(1) (2016), 5–12.
[36] T. Poteat, K. Rachlin, S. Lare, A. Janssen and A. Devor, History and Prevalence of Gender Dysphoria, in: Transgender Medicine, Springer, 2019, pp. 1–24.
[37] A. Suess Schwend, S. Winter, Z. Chiam, A. Smiley and M. Cabral Grinspan, Depathologising gender diversity in childhood in the process of ICD revision and reform, Global Public Health 13(11) (2018), 1585–1598.
[38] L. J. Esserman, I. M. Thompson, B. Reid, P. Nelson, D. F. Ransohoff, H. G. Welch, S. Hwang, D. A. Berry, K. W. Kinzler, W. C. Black, M. Bissell, H. Parnes and S. Srivastava, Addressing overdiagnosis and overtreatment in cancer: a prescription for change, Lancet Oncol. 15(6) (2014), e234–e242.
[39] A. Synnott, The body social, Routledge, 2002.
[40] B. Hofmann, Limits to human enhancement: nature, disease, therapy or betterment, BMC Med. Ethics 18(56) (2017), 1–11.
[41] B. Hofmann, On the dynamics of sickness in work absence, in: Social aspects of illness, disease and sickness absence, H. Nordby, R. Rønning and G. Tellnes, eds, Unipub forlag, Oslo, 2011, p. 287.
[42] B. Hofmann, On the triad disease, illness and sickness, J. Med. Philos. 27(6) (2002), 651–674.
[43] K. D. Clouser, C. M. Culver and B. Gert, Malady, in: What is a disease? R. F. Almeder and J. M. Humber, eds, Humana Press, Totowa NJ, 1997, pp. 173–217.
[44] B. Hofmann, Perspectives on human malady: disease, illness, and sickness, in: The Routledge Companion to Philosophy of Medicine, M. Solomon, J. Simon and H. Kincaid, eds, Routledge, London, 2016.
[45] I. Hacking, Mad Travelers: Reflections on the Reality of Transient Mental Illnesses, University Press of Virginia, Charlottesville, 1998.
[46] R. Mayes and A. V. Horwitz, DSM-III and the revolution in the classification of mental illness, J. Hist. Behav. Sci. 41(3) (2005), 249–267.
[47] H. Helmchen, M. M. Baltes, B. Geiselmann, S. Kanowski, M. Linden, F. M. Reischies, M. Wagner, T. Wernicke and H.-U. Wilms, Psychiatric illnesses in old age, in: The Berlin Aging Study: Aging from 70 to 100, 1999, pp. 167–196.
[48] D. Bolton, What is mental disorder?: an essay in philosophy science, and values, Oxford University Press, 2008.

[49] R. Cooper, Diagnosing the diagnostic and statistical manual of mental disorders, Routledge, 2018.
[50] D. J. Stein, K. A. Phillips, D. Bolton, K. W. Fulford, J. Z. Sadler and K. S. Kendler, What is a mental/psychiatric disorder? From DSM-IV to DSM-V, Psychol. Med. 40(11) (2010), 1759–1765.
[51] J. C. Wakefield, Disorder as harmful dysfunction: a conceptual critique of DSM-III-R's definition of mental disorder, Psychol. Rev. 99(2) (1992), 232–247.
[52] T. Burge, Philosophy of Language and Mind: 1950--1990, Philos. Rev. 101(1) (1991), 3–51.
[53] T. Burge, Individualism and the Mental, in: The Twin Earth Chronicles, A. Pessin and S. Goldberg, eds, M.E. Sharpe, New York, 1996, pp. 125–142.
[54] A. Mol, The body multiple: Ontology in medical practice, Duke University Press, 2002.
[55] R. E. Kendell, The role of diagnosis in psychiatry, Blackwell Scientific Publications, Oxford, 1975.
[56] K. Jaspers, Die Begriffe Gesundheit und Krankheit, in: Allgemeine Psychopathologie, Springer, Berline, 1913/1973, pp. 651–711.
[57] C. Boorse, A Rebuttal on Health, in: What Is Disease? J. Humber and K. Almeder, eds, Humana Press, Totowa, New Jersey, 1993.
[58] W. Stempsey, Emerging medical technologies and emerging conceptions of health, Theor. Med. Bioethics 27 (2006), 227–234.
[59] H. T. Engelhardt and K. W. Wildes, Health and Disease – Philosophical Perspectives, in: Encyclopedia of bioethics, W. T. Reich, ed., MacMillan, New York, 1995, pp. 1101–1106.
[60] W. E. Stempsey, Disease and Diagnosis: Value-dependant Realism, Kluwer, Dordrecht, 1999.
[61] B. Hofmann, On the triad disease, illness and sickness, J. Med. Philos. 27(6) (2002), 651–674.
[62] B. Hofmann, Complexity of the concept of disease as shown through rival theoretical frameworks, Theor. Med. Bioethics 22(3) (2001), 211–237.
[63] R. L. Engle and B. J. Davis, Medical diagnosis: present, past, and future. I. Present concepts of the meaning and limitations of medical diagnosis, Achs. Inter. Med. 112 (1963), 512–519.
[64] H. G. Wells, The Classificatory Assumption, in: First and Last Things, Constable, London, 1908.
[65] H. Nordby, Vague concept of disease in the physician-patient relations, Tidsskr. Nor. Laegeforen. 125(6) (2005), 765–766.
[66] K. Sadegh-Zadeh, Fuzzy health, illness, and disease, J. Med. Philos. 25(5) (2000), 605–638.
[67] R. Seising, From vagueness in medical thought to the foundations of fuzzy reasoning in medical diagnosis, Artif. Intell. Med. 38(3) (2006), 237–256.
[68] R. A. Aronowitz, The converged experience of risk and disease, Milbank Q. 87(2) (2009), 417–442.
[69] R. M. Kaplan, Disease, Diagnoses, and Dollars: Facing the Ever-Expanding Market for Medical Care, Springer Science & Business Media, 2009.
[70] R. Moynihan, Drug firms hype disease as sales ploy, industry chief claims, BMJ 324(7342) (2002), 867.
[71] R. Moynihan, J. Brodersen, I. Heath, M. Johansson, T. Kuehlein, S. Minué-Lorenzo, H. Petursson, M. Pizzanelli, S. Reventlow and J. Sigurdsson, Reforming disease definitions: a new primary care led, people-centred approach, BMJ evidence-based medicine, bmjebm-2018-111148 (2019).
[72] R. Moynihan, I. Heath and D. Henry, Selling sickness: the pharmaceutical industry and disease mongering, BMJ 324(7342) (2002), 886–891.
[73] B. Hofmann, Expanding disease and undermining the ethos of medicine, Eur. J. Epidemiol. 34(7) (2019), 613–619.
[74] P. Conrad, The medicalization of society: On the transformation of human conditions into treatable disorders, Johns Hopkins University Press, Baltimore, 2007.
[75] H. G. Welch, L. Schwartz and S. Woloshin, Overdiagnosed: making people sick in the pursuit of health, Beacon Press, Boston, Mass., 2011, xvii, 228 p.

[76] B. M. Hofmann, Expanding disease and undermining the ethos of medicine, Eur. J. Epidemiol. 34 (2019), 613–619.
[77] B. Hofmann, The technological invention of disease – on disease, technology and values. University of Oslo, 2002.
[78] A. Caplan, How Can Aging Be Thought of as Anything Other Than a Disease? in: Handbook of the Philosophy of Medicine, T. Schramme and S. Edwards, eds, Springer, Netherlands, Dordrecht, 2017, pp. 233–240.
[79] B. Hofmann, Obesity as a Socially Defined Disease: Philosophical Considerations and Implications for Policy and Care, Health Care Anal. 24(1) (2016), 86–100.
[80] M. C. Burke, Resisting pathology: GID and the contested terrain of diagnosis in the transgender rights movement, Adv. Med. Sociol. 12 (2011), 183–210.
[81] R. P. Harris, Invited commentary: beyond overdiagnosis—diagnosis without benefit, Am. J. Epidemiol. 188(10) (2019), 1818–1820.
[82] R. P. Harris, S. L. Sheridan, C. L. Lewis, C. Barclay, M. B. Vu, C. E. Kistler, C. E. Golin, J. T. DeFrank and N. T. Brewer, The harms of screening: a proposed taxonomy and application to lung cancer screening, JAMA Int. Med. 174(2) (2014), 281–286.
[83] M. B. Strauss, Familiar medical quotations, in: Familiar medical quotations, Little, Brown and Company, 1968.
[84] J. Y. Halpern, Reasoning about uncertainty, MIT press, 2017.
[85] S. Hatch, Uncertainty in medicine, BMJ 357 (2017).
[86] S. O. Hansson, Decision making under great uncertainty, Philos. Soc. Sci. 26(3) (1996), 369–386.
[87] P. K. J. Han, W. M. P. Klein and N. K. Arora, Varieties of uncertainty in health care: a conceptual taxonomy, Med. Decis. Mak. 31(6) (2011), 828–838.
[88] The Norwegian Medical Association. Gjør kloke valg – Radiologi Oslo: The Norwegian Medical Association, 2018 [Available from: https://www.legeforeningen.no/foreningsledd/fagmed/norsk-radiologisk-forening/artikler/fag-og-utdanningsstoff-fra-noraforum/gjor-kloke-valg-radiologi/].
[89] B. Hofmann and S. Holm, Philosophy of Science, in: Research in Medical and Biological Sciences: From Planning and Preparation ot Grant Application and Publication, P. Laake, H. B. Benestad and B. R. Olsen, eds, Academic Press, 2015, pp. 1–42.
[90] J. Stegenga, Medical nihilism, Oxford University Press, 2018.
[91] T. Merricks, Varieties of vagueness, Philos. Phenomenol. Res. 62(1) (2001), 145–157.
[92] G. Keil, L. Keuck and R. Hauswald, Vagueness in psychiatry, Oxford University Press, Oxford, 2017.
[93] P. Hucklenbroich, Disease entities and the borderline between health and disease: Where is the place of gradations, in: Vagueness in psychiatry, 2017, pp. 75–92.
[94] C. G. Hempel, Science and human values, 1965.
[95] H. Douglas, Why inductive risk requires values in science, in: Current controversies in values and science, 2017, pp. 81–93.
[96] T. Wilholt, Bias and values in scientific research, Stud. Hist. Philos. Sci. 40(1) (2009), 92–101.
[97] A. Plutynski, Safe or Sorry? Cancer Screening and Inductive Risk, in: Exploring Inductive Risk: Case Studies of Values in Science, 2017, p. 149.
[98] H. Douglas, Values in science, in: The Oxford handbook of philosophy of science, 2016, pp. 609–632.
[99] J. B. Biddle, Inductive risk, epistemic risk, and overdiagnosis of disease, Perspect. Sci. 24(2) (2016), 192–205.
[100] B. Hofmann, K. Bond and L. Sandman, Evaluating facts and facting evaluations: On the fact-value relationship in HTA, J. Eval. Clin. Pract. 24(5) (2018), 957–965.

[101] B. Hofmann, I. Cleemput, K. Bond, T. Krones, S. Droste, D. Sacchini and W. Oortwijn, Revealing and acknowledging value judgments in health technology assessment, Int. J. Technol. Assess. Health Care 30(6) (2014), 579–586.
[102] B. Williams, Ethics and the Limits of Philosophy, Harvard University Press, Cambridge, MA, 1985.
[103] B. Williams, Ethics and the Limits of Philosophy, Taylor & Francis, 2011.
[104] R. Stoecker and G. Keil, Disease as a vague and thick cluster concept, in: Vagueness in Psychiatry, 2016.
[105] K. Van Deemter, Not exactly: In praise of vagueness, Oxford University Press, 2012.
[106] M. Kwiatkowska, The Beauty of Vagueness, in: On Fuzziness, Springer, 2013, pp. 349–352.
[107] B. Hofmann, The Paradox of Health Care, Health Care Anal. 9 (2001), 369–383.

Fabio Krykhtine

8 Vagueness in technology

Abstract: We give a discussion and overview of vagueness, concepts and accepted knowledge, that gives a better understanding between its philosophical implications and the actual engineering and current computational applications. The discussion considers education development, the limits of computation, and a modern way to deal with vagueness in present day high technology. In addition, we show some of the tools that can be used to interact with vagueness and illustrate, by means of short practical examples, how vagueness burdens and limits classification and hierarchization, decision making, and artificial intelligence. Also, we suggest a research roadmap for broadening our knowledge and improving technology while, at the same time, we highlight the challenges involved.

8.1 Introduction

During the last century, in terms of technology, the world had faced intense changes that transformed the life and the future of humans and the concept of environment. It was a phenomenon that had a strong impact on the planet that was caused by the extensive use of natural resources in a very accelerated industrial process that had to meet the mass production and a large portfolio of network technologies applications. It involves not only products, but services and the view of competition in a global sense [1].

Science had finally been utilized since a great part of the knowledge produced during the last centuries has finally found practical applications. In particular, knowledge produced in many fields of science had been utilized to develop equipment and processes that created, in turn, new products and services.

In the fields of mathematics and electronics, the Turing machine is considered the most important idea that has had a very deep impact. This conceptual machine aggregated the possibility to perform many computational tasks by introducing the concept of automatic or mechanical computation. From its inception in 1936 [2], the machine inspired engineers to develop real computers and gradually transform their slow devices with limited capabilities into ultrafast machines with virtually unlimited storage and the capability to operate in parallel as nodes of a global computational network. This development is clearly a big step forward as it allowed many human activities to "operate" in a collaborative mode. In addition, supplies, logistics, production, communications, finance, etc., is now organized by engineering sciences, information technology, and process designs. All of those great developments made

Acknowledgement: I would like to thank Apostolos Syropoulos for his help.

https://doi.org/10.1515/9783110704303-008

the last century incredible. Thanks to these developments, we now have better automation, control, and management.

Despite this success, it causes much annoyance that an important idea that played a crucial role in it, is really ignored. Unfortunately, the technological avalanche caused by the global industry, which has made its aim to satisfy every possible need, artificial or real, by trying to go beyond existing technological limits, does not acknowledge the exploitation of *vagueness*. Some called it *uncertainty* but I prefer to call it vagueness.

Vagueness is everywhere in the real world. In particular, humans operate in a vague environment in a vague way. However, most engineers and scientists are not feeling really comfortable with vagueness so they try to oust it from their "world" and replace it with lack of *precision*. Of course, this happens because the mathematical apparatus we use assumes we live in a precise world and all "fluctuations" can be perfectly described with statistical science, which most people understand very well.

Uncertainty was generally attached to the idea of risk existence, to unpredictable and misunderstood situations, instability, and everything else that gets in the way of a project, a plan, or a forecast, wrongly associated with an unstable and "fuzzy" situation. So, precision is something that characterizes a good and well-thought-out plan or project, whereas uncertainty and lack of ways to handle it, is an indication, if not a proof, that a plan or a project is not well-thought-out and bound to fail.

When we have to fit a nut to a screw, turn a part, or even produce a gear for a Swiss watch mechanism, we witness the idea of precision and its usefulness; nevertheless, even there is imprecision. Uncertainty lives there but it does affect the reputation of the state of art of a Swiss watch. Indeed, what gave the reputation to the Swiss watches was their extreme precision mechanism and, of course, the idea behind it. During the 1970s, the introduction of the quartz watch shook the watch manufacturers, putting the Japanese industry ahead, promoting the extinction of many of the world's brands. Clearly, the concept of precision and imprecision are relative. Although the quartz watch is much more precise than its analog counterpart, still it goes slow by 10 seconds every month.

Without the widespread idea of precision, a series of relationships could be impossible, especially at the industrial production sector where standardization has taken place in assembly lines for mass products. It is important to join many parts, providing enterprises to work in a global linked supply production system that is a common reality in the current global industrial environment.

For a long time, uncertainty was studied only in philosophy and humanities where it was considered something completely subjective. It provided a good starting point for discussions about the possibility to use the concept of uncertainty and the lack of information as part of our everyday analyses. Many scientists and philosophers such as Jan Łukasiewicz [3], Bertrand Russel[1] [4], Max Black [5], Lotfi Zadeh [6], and others, even artists such as Henri Matisse, had expressed a disquiet about uncertainty

1 His full name was Bertrand Arthur William Russell, 3rd Earl Russell.

and how models of the world that did not take under consideration uncertainty were unable to provide a faithful representation of reality.

Understanding different people, cultures and minds, behavior, recognizing patterns and personalities, asking questions, answering questions, and more, interpreting information, is not possible in a bivalent way, and a yes–no fashion, we need to embrace uncertainty to fully understand these concepts and ideas.

Uncertainty is something that must be totally explored with creativity and an open mind by science. Exploring the incredible universe of possibilities and categorizing the different forms of uncertainties that are used to express the wealth of information we process every single day (e. g., vagueness, imprecision,imprecision ambiguity, lack of information, etc.) are components of a great unexplored glossary. In particular, when studying the concept of imprecision, we encounter a vast number of concepts linked to uncertainty and imprecision that were explored by Mark J. Wierman [7].

When ordinary people started to interact with computers by using with particularly difficult programming languages, it was quite evident that we had to cover a great distance in order to make the communication between humans and computers more human-like. However, this, let us call it a linguistic and methodological gap, allowed a relatively small number of people to develop a community of weirdos that knew how to program computers and made their business a huge success by creating corporations specialized in consulting, services, and technology supplies. On the other hand, since languages are intrinsically vague, computer and system designers have to adapt their design philosophies so to take under consideration this fundamental feature of natural languages. A way to solve this problem is to leave the mass process model of development and to adopt a model based on a mass customization process. A side effect of this model is a better usage of natural resources and raw material, decrease of carbon emissions, and emphasis on renewability. More generally, this approach can benefit companies outside the IT sector. For example, any company has to use a flexible system that is capable to predict changes in markets, adapt to competition, produce better mass production strategies, control demand, and be assertive during maintenance periods. And of course, any company capable of dealing with vagueness should be at least one step ahead from its competitors who do not understand it.

8.2 Uncertainty from theory to practice

Engineers face real world problems that demand their constant attention so to deliver the proper response to any challenge posed by contractors. In this process, they develop new technology that has to operate in an environment that is characterized by imprecision, intolerance, and nondiscrete values. At the same time, this environment is usually mathematically described in a way that completely ignores these characteristics.

When an engineer has to make a prediction for an unstable process, to perform an evaluation of a multicriteria decision problem, to organize collaborative decision environments, the utilization of methods that have uncertainty and vagueness as important components is more than necessary! Such methods allow us to use data in a better way in computation [8]. In addition, the opinions and/or solutions proposed by specialists are quite vague and the information available for the solution of a problem are not written down in a rigorous way. Instead they are utterances that cannot be easily modeled using ordinary mathematics.

Unfortunately, the situation just described is quite awkward for most professionals and students that had never learned about how to explore uncertainty. Clearly, this happens because there is a resistance to accept that uncertainty, imprecision, and lack of information can be "tamed" in computation processes.

Our "habit" to classify things in a bivalent way (e. g., with a yes or a no) is something we learn early in our lives, and unfortunately, when we grew up we fail to see things in the middle. However, the world does not behave this way and many cases are boundary cases, that is, cases where it is not clear if a yes or a no can clearly classify something. Unfortunately, it is not easy to learn to think in a different way although psychiatry tells us that adults usually are able to think in shades of gray. This forms some sort of oxymoron but the real problem is clearly what is wrong is the way we have been taught to perceive the world.

Almost every single day we have to make some complex decision and if could analyze the way we make such decisions we would be surprised by the way our brains make selections and processes data. For example, for any human being a temperature is not a number but a physical sensation (e. g., cold, very cold, hot, absolutely hot, etc.). When we associate a scalar value to a temperature, we actually interpret a physical sensation by it. More general, perception can be associated to linguistic terms, words, colors, and also numbers. The act of collecting and registering perceptions can be done in many ways using qualitative information, which is also subjective something we need to take under consideration when evaluating something. In summary, to process information the way computers do is wrong!

This entrenchment of the mental concept to a binary way is so imperative that it limits people's cognitive ability to understand and accept intermediate grade states. As result, it affects social, political, economical, environmental, and all human relations making room for analysis by intolerance. People should be educated to understand that not everything is divided into a duality but, in the vast majority of cases, in a set of intermediate states between clearly distinct boundaries.

So, as a consequence, our minds learned a nonnatural set theory, which is more connected to antique machine processing and, in a adult age, we present resistance to understand different theories approaches. In other words, thinking in a binary way is adequate to computers because that is the way it was invented.

That's why most people express difficulties to accept uncertainty and are closed to accept this knowledge.

Imagine if the "tools for vagueness" had been exhausted developed before the Turing machine. In resume, we should make computers more similar to human, not humans more similar to computer. It involves two sides, reformulate computer cores by introducing other computations technologies, and review human education curriculum, especially when presenting sets theory [9].

When engineers have to work in a vagueness territory, most of then are not prepared enough because they are rooted to a discrete mathematics mode, considered crisp. They also have as tools a group of methods sufficiently presented at academic times, supported by statistic and crisp mathematical models. Engineers rarely learn to deal with vagueness, uncertainty, and imprecision at the graduation course, and also rarely approach a project by the point of view of uncertainty, when they are bent on project desks.

Vagueness, uncertainty, and imprecision have taken part in discussion for a long time, but especially in the 20th century, philosophers, scientists, and researchers began to dedicate time to explore the concepts to better describe uncertainty, imprecision, vagueness, fuzziness, etc.

Jan Łukasiewicz introduced in 1917 the development of three-valued propositional calculus published as a paper titled "On Three-valued Logic" in 1920. It was one of the first initiatives and it received recognition that there was a need for intermediate values between boundaries. Łukasiewicz assumed, in 1919, as the Polish Minister of Education, proposed a curriculum that emphasized the early acquisition of logical and mathematical concepts.

Bertrand Russell, expressed the necessity to develop a logic more adherent to the reality. Russell's work "Vagueness" is a standard reference for many works. Russell declared those sets and placed together on it groups on his work that the vagueness has degrees. Russell plays: "I propose to prove that all language is vague and, therefore, my language is vague, but I do not wish this conclusion to be the one that you could derive without the help of the syllogism. I shall be as little vague as I know how to be if I am to employ the English language."

A name that must be highlighted as an important contribution in the exploration of the uncertainty and imprecision in the universe is Max Black. Black wrote a paper entitled "Vagueness: An exercise in logical analysis" in 1937 and developed works attached to metaphors, linguistics, and mathematics. Black developed a little bit beyond on degrees of vagueness. His work has relevance to the development of a new frontier of uncertainty processing with discussion. His development about metaphors is a view to better understand meanings, cultural differences, and linguistic expression.

In 1965, Lotfi Zadeh [6] published his paper that introduced fuzzy set theory, and finally introduced the membership function as a way to declare the degree of true (or presence) that associates a level from an element to one or more sets. Zadeh had not only explored the entire concept but also organized the whole theory that gives basics to the modern fuzzy sets theory addressing the tools necessary to process imprecise information as the human brain does.

In chronological order, it was presented 30 years after the computational model proposed by Alan Mathison Turing [2]. Zadeh had a hard time even publishing his work. The world was, at that moment, engaged in the development of computers that were more and more advanced, but based on binary processing. The fuzzy set theory did not have any attention at that time, especially in the western world.

In the other hand, in the oriental world, especially at Japan, during the 1970s and the 1980s, Japanese scientists discovered applications attached to machines [10], using rules and on-board decision making to run simple processes of automation, launching the first intelligent machines dedicated to mass consumer markets, ranging from small machines such as refrigerators, washing machines, rice cookers, auto focus cameras, and more efficient car brakes' systems. The products were processing data according to sensors and following rules to perform actions. Japan introduced applications in control and automation, with processes attached to water treatment plants [11] and control for urban trains' automation.

Since the first application utilizing rules in a format of sentences, the intelligent systems had a brutal advance. The Mamdami [12] rules were initially proposed to control a steam machine harmonizing temperature and pressure. These rules are named after Ebrahim H. Mamdani.

When Mamdami rules [13] were embedded on a processing chip, the Japanese industries took a big step that positioned its products with the image of the most modern in the world, being quite competitive and preferred by consumers all over the planet. It had registered in the 1980s nearly 2000 patents involving intelligent machines. It was a wide field for developing new products and offered new goodies to markets.

From intelligent machines, the industry followed the road map to intelligent systems, where many machines are integrated in systems. The systems are composed by many algorithms that work together in a chain where some provide information to others in a network. Some of these systems also have the capability to optimize itself and dynamically upgrade levels of knowledge.

It can be seen in smart cities [14] and IoT technology, for example, in smart traffic control. By monitoring cameras and even vehicles by internet connectivity, the synchronicity of traffic signal can be adjusted to provide better flows. It consists of a set of rules and a set of inputs, provided by different sensors and nodes that adjusts itself to a considered optimum net performance.

The first and much important tip to interact with vagueness, imprecision, and uncertainty is to accept the world with it enormous and natural imprecision. For the most part for the person, it is very difficult. Recognizing imprecision is a fundamental point to start to work with it. Especially in engineering, the classical trend is to work, project, decide, and build by calculation running over discreet mathematics.

For a long time, researchers and scientists have pursued precision and certainty as a way of developing operations, especially in the case of numerical operations. Scalar quantities are associated with a precision, obtained by a tolerance level sufficiently acceptable for real operations.

Dialogue, even unconsciously, with the tolerance for imprecision, in which we adjust the precision to our convenience is a key point. It is a fantastic economy of time and computation costs. The uncertainty and the precision are concepts connected by costs. Those costs can appear in collecting data, in processing data or in delivering data phases. It took time, energy and at a minimum, an investment. That's why the precision and the tolerance is a relevant aspect to categorize the kind of solution that will be proposed.

In computation, the cost dilemma occurs when a complex system is solved by brutal computation force. It is a large cost to obtain responses in a short time or a long processing period to obtain the same. The solution can wait. Do we need extreme precision or can our needs be more tolerant? Sometimes it is out of scope or impossible to be solved in relevant time.

Those questions should be done when defining the computation method to work with complex uncertainty. The method to deal with limited computation power is to work with optimized processing systems and avoid brutal force computation caused by exponential explosions in scenarios modeling. Soft computation appears as a good candidate to solve or reduce the computational burden.

A company's accounting system can accept a margin of error for cents or a speed camera can accept the variation of some units when registering a traffic ticket. In this sense, we are dealing with measures of real cases that can accept flexibility in their results. But what would be the limit of this flexibility?

To make a decision, a response such as: "this one is much better than that other" can be enough for most of the cases.

Timothy J. Ross [9] and his colleagues advocated that fuzzy logic can be applied in two situations: 'When it is very complex and faces a nonunderstood behavior, and in situations when an approximate result, is delivered in a very fast way is required."

In modern engineering, the mastery of new instruments that allow dialogue with flexibility and uncertainty has become essential to portray a reality more similar to the actual expression of a phenomenon, but in addition, to allow measurement by descriptions that are sufficient in their linguistic significance to represent a phenomenon or a technical need. Then the universe of possibilities expands when imprecision starts to be recognized, accepted, and worked in a dialogue with a described world more adherent to reality, as a way of establishing freedom for science, especially for engineering, that starts to work with greater creativity to provide the most appropriate solution of real-world problems.

The possibility of engineering to work with imprecision, in turn, opened the possibility to work with more complex and flexible models, accepting input variables containing different margins and types of uncertainty. This approach allowed an expansion of the capacity of automation, control, decision making, prediction, etc., through a change of paradigms that moved the engineering of a mathematical world based on exact, discrete, and improperly precise mathematics models to a format that computes information from different typologies, origins, and sources, including processing by linguistic variables, intervals, symbols, and numerical inputs.

The complexity involving inexact measures and delivering a calculation output made researchers develop new methods and tools, observing the divergences between the preconceived models and the reality. The computational limits affected many solutions and new development [15]. Once many kinds of variables could be processed as inputs and outputs; the computation possibilities expand for new limits in a more far frontier.

Nowadays, with the technology development in computational power and processing multisources in computation, machines, and systems become more efficient applying intelligence and autonomous processing. These facts made possible a big turn in technologies and the possibility that made calculus with uncertainty from the input to the output.

Zadeh has mapped out four types of fuzzy algorithms:
Definitional that categorize fuzzy input.
Generational that create fuzzy output.
Relational that describe systems flux.
Decisional that issue commands from ongoing feedback.

8.3 Thinking from simple to complex

8.3.1 Case 1 – classifying apples

Consider the very simple task of classification apples contained in a box that contains 18 Kg of mixed size apples [16]. If all of the apples are arranged in a table and examined carefully, when the separation task starts, the elements that belong to the biggest apple group will be found easily and even identified as the biggest apples (maximum of maximum set) and the smallest apple (minimum of minimum set) from the set of apples. Now, it should be obvious that the biggest and the smallest apples set the upper and the lower bound, and consequently, we will classify the rest of the apples using a similar approach.

Let us try to divide them into two groups: the big apples and the small apples. Each apple will be compared to members of each group and will be placed to the group where it belongs. At the beginning, this process is easy but as we proceed we will find apples that fit perfectly to both groups. Naturally, the question is: To which group does this apple belong?

In most cases, the decision is quite subjective but in a number of cases we may deduce that these apples belong to both groups! Thus one may wonder whether we should create a third group that would contain all middle-sized apples. Even we create this third group of apples, at some point we will find an apple that does not fit in any group. However, it is possible to solve this problem by putting a strict limit (e. g., any apple that weighs above 100 gr belongs to a certain group). This is an idea that is known in the literature of fuzzy sets as α-cut.

In this example, we have encountered a good number of ideas and concepts that are quite relevant to fuzzy set theory: maximum, minimum, fuzzy boundaries, α-cuts, etc. Using these ideas, we could devise a separation algorithm that could sort the apples by assigning a membership degree to each apple. Then our task would become much more easy. Figure 1 illustrates this idea.

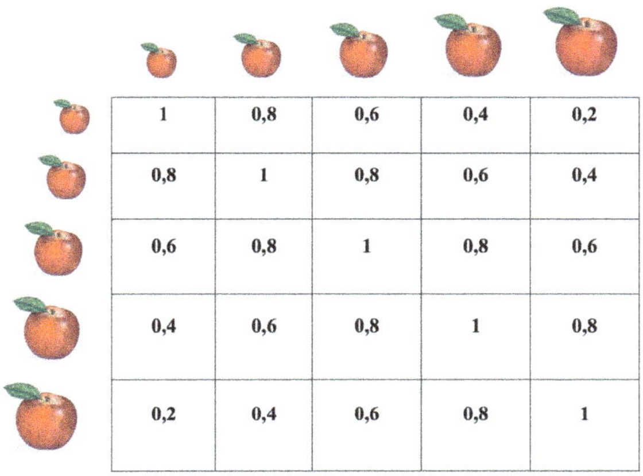

1	0,8	0,6	0,4	0,2
0,8	1	0,8	0,6	0,4
0,6	0,8	1	0,8	0,6
0,4	0,6	0,8	1	0,8
0,2	0,4	0,6	0,8	1

Figure 1: Apples and membership degrees.

8.3.2 Case 2 – rockets and toys

An amazing engineering world famous case of complex operation regarding control and automation is the reusable rocket engines landing. Its is extremely impressive the way a rocket lands, bringing the entire expensive equipment back, reducing costs, and providing more frequently space missions.

It is a case where the controlling stability systems [17] returns result for adjusting the rocket trajectory in very fast responses. In terms of requirement, that real-time reaction, very specific and fast enough is capable to land the rocket in a vessel, even with the sea movement affecting the vessel floating. The project failed many times until we obtained adjusted success; this is normal in developing research projects.

The same technology that provides a rocket return to a vessel is applied in toys as small two-wheel robots with capability to compensate forces opposite to its equilibrium point through executing adjustments to its traction system, even if you add mass on it arms affecting the central gravity point.

In a rocket, compared to the robot toy, the third axis is included and a path involving the expected trajectory to the mission arrival point is plotted as the reference.

Sensors are placed at the rocket and at the vessel transmitting inputs to a processor that return outputs to aerodynamic guidance, auxiliary engines, and land gear.

It works based on measuring the divergences and returning as a response and opposite axis force, adjusted to the intensity of the divergence that adjust the element to the central axis. For example, the operation of the guidance and landing gear are determined by the following rules [13].

GUIDANCE CONTROL RULES

IF the *X* axis divergence is "small"
THEN apply to the *X*-axis an "opposite small force"
ELSE wait 1/1000 seconds.
IF the *Y*-axis divergence is "small"
THEN apply to the *Y*-axis an "opposite small force"
ELSE wait 1/1000 seconds.
IF the *Z*-axis divergence is "small"
THEN apply to the *Z*-axis an "opposite small force"
ELSE wait 1/1000 seconds.
IF the *X*-axis divergence is "medium"
THEN apply to the *X*-axis an "opposite medium force"
ELSE wait 1/1000 seconds.
IF the *Y* axis divergence is "medium"
THEN apply to the *Y*-axis an "opposite medium force"
ELSE wait 1/1000 seconds.
IF the *Z* axis divergence is "medium"
THEN apply to the *Z*-axis an "opposite medium force"
ELSE wait 1/1000 seconds.
IF the *X* axis divergence is "big"
THEN apply to the *X*-axis an "opposite big force"
ELSE wait 1/1000 seconds.
IF the *Y* axis divergence is "big"
THEN apply to the *Y*-axis an "opposite big force"
ELSE wait 1/1000 seconds.
IF the *Z* axis divergence is "big"
THEN apply to the *Z*-axis an "opposite big force"
ELSE wait 1/1000 seconds.

LANDING GEAR CONTROL RULES

IF the mission arrival point is "far"
THEN keep "landing gear closed"
ELSE wait 1/1000 seconds.
IF the mission arrival point is "near"
THEN keep "landing gear open."

By these sets of rules and comparing to expected trajectory prediction, the rocket returns to the vessel. Of course, it is being treated in a simplified way just to illustrate that it is a new application to an old algorithm that is very well known [18]. The central point here is to show that the rules are linguistic and it values, expressing what means "far" or "near" to a rocket, are addressed by fuzzy numbers.

The rocket landing could be a reality because today the state of art has a computation method and computation power that merges sensors, processing, and physical movement responses with enough time to execute the task. As an engineering application, it is impressive, making investors believe and verify that it is the edge technology representing a promise of return to the capital invested.

8.3.3 Case 3 – Location studies and economic issues

The imprecision in engineering sometimes inserts professionals in tasks to predict scenarios, project an artefact without enough information, or take decisions in a vast and complex environment.

Especially in industrial engineering, when facing location studies for industrial plants or at economic sciences, which occurs a great deal with data provided sometimes full of uncertainty, working with tools that consider uncertainty existence are important to achieve good enough results.

Carlos Alberto Nunes Cosenza, an emeritus senior professor at Federal University of Rio de Janeiro, developed during the last four decades a method for evaluating placing fuzzy sets, linguistic values, and informant/specialist data processing.

It is called the COPPE COSENZA MODEL [19] and was created based on his large experience as an economist and industrial engineer in organized and planned territory polices [20]. Cosenza presented the COPPE COSENZA MODEL for location sites and opened an avenue for a specific algorithm to deal with the uncertainty.

For him, life is composed by two matrices: need and attendance. By crossing these two matrices, he found a third matrix: the relationship matrix.

In location studies, the industry characteristic (need matrix) is considered in a demand matrix (requisition) that must contain the industry necessities described in terms of requisites for natural and infrastructural factors levels. On the other side, a matrix for supply (attendance) its necessities is composed, each one representing a candidate place to receive the scope projected industrial plant.

The solution is an interaction between the two matrices and the chosen candidate is the one that best fit the demand. In these scenarios, Cosenza applied the third matrix to adjust relationship between the supply and demand matrix because he had an interesting insight about abstract matrix.

He comprehended, in practice, the effects of information lack and uncertainty during his work life trajectory and had create a group of tools to overcome it providing a more intuitive processing of data.

If is considered a zero value in the supply, does it means the same of a zero value at the demand? Applying to a real world context, null, zero and absence signify the same thing?

That discussion was very rich and took place when computation tools faced data with classical approach presenting disturbs on it results, specially when the zero was multiplied at factors list.

It opens space for a philosophical discussion where the absences are different and it is addressed to the point where crisp mathematics meets limitations. It is simple, but not obvious to understand, by point of view of real world application. It surges when a very experienced professional questing the machine about it limitations and designing a tool adapting the limited machine to the task, by inserting a set of new operators trough the relation matrix.

A score a computation can not penalize a site candidate location because the supply matrix contains what is not needed in a demand matrix. Simple as that, a zero at traditional approaches can prejudice a lot of situations in mathematics while, in the reality, it received completely different meanings, depending where it is placed. A zero at Demand or a Zero at the Supply matrix?

Cosenza named absences differences: A zero means a value for a level. A null set represents that these set is not considered. An empty set means an opportunity, once if is being considered infrastructural factors: as roads, transmission lines, IT; that can be installed, transforming the candidate site in a high level candidate by implementing some level of investment. If a zero is placed at supply matrix and the factor is needed by the demand matrix, the supply to this factor is inferior. If a zero is placed at demand matrix and the supply matrix presents a value, the offer for the factor is superior.

The Coppe Cosenza model was finally improved at Bio-diesel Brazilian Program headed by Petrobras for location studies considering a set of five different industrial plant types [21]. As a challenge for processing, the Brazilian Bio-diesel Program offered 1789 candidate cities sites for receiving investments and organize a complex logistic and industrial production arrangement, making use of the existing ethanol routes, ports, agricultural, and demographic aspects, covering an territory biggest than France, Spain, Germany, and Italy together.

The model had presented all the results in a Geographic Information System (GIS) platform and exhibit the group of interventions in infrastructure that could made the scenario indeed better. It was the concept of the empty sets in practice showing how an increase in the infrastructure would provide better returns, transforming some areas evaluated in a good rate, into an excellent rate, by simulating effects of investment in a regional territory. The location model follows the specialist mindsets and it rules to merge all data, opinions, and guidelines according to the project nature and all specific characteristics. By the specialists, all matrices are fed and the factors are categorized in levels converted to linguistic values. The factors are divided in specific and general factors and classified in levels of importance such as: irrelevant, little conditionant, conditionant, or crucial.

The COPPE COSENZA MODEL was also used by the Brazilian Nuclear Program to investigate candidate sites for new nuclear power plants, helping to draw an strategic implementation map that observes environmental, social, and economical aspects.

Recently, Cosenza proposed the application to cross data from population health and industrial activities developed at some sites and showed groups of diseases promoted by the long exposition to some of the industrial wastes.

It is very clear today that COPPE COSENZA MODEL is a huge tool for uncertainty processing, applied in many science fields such as wealth, fashion industry, educational evaluation, aviation, and planning, but as in all disruptive ideas it faced some opposition along the first years.

Once a time, in seminar at Cambridge, along his academic life, in a occasion when visitors present their developments, Cosenza was invited to talk about his work. One of the gentle organizers read his abstract and named his presentation as "Difference between Zeros."

He was motivated and explaining his point of view while one of the persons in the audience said to him that his work had no fundamental. The person criticized him saying in good volume: "Zero is zero and nothing is nothing." Cosenza gently struck back to the colleague provocation that he does not believe that the zeros should be treated in different ways.

With this situation, Cosenza asked the colleague if he had any enemies in his life; someone he could not stand near. The colleague answered: "YES," and then, Cosenza asked if the enemy was in the audience group, and the person answered "NO." Cosenza wrote on the blackboard "ENEMY = ZERO."

In the sequence, Cosenza asked to the colleague if exists someone that he loves very much, someone very special to him, and the colleague answered: "YES." Again, Cosenza asked: "Is she here?" and the colleague answered "NO." Cosenza wrote on the blackboard "LOVE = ZERO."

After that, Cosenza pointed to the sentences on the blackboard: "ENEMY = ZERO" and "LOVE = ZERO," and asked the person: "Do these two absences mean the same for you"?

Cosenza tried to explain to the colleague that the zero is constantly addressed to absence, to neutral value, and to null value, but in different sciences, as economics, this approach should be questioned rather than simply applied with the use of crisp mathematics. It is in another context in which abstract matrices and computational tools appropriate to the need and reality of the calculation that must be used, in order not to distort the desired result. It is as if the same tooling was used for everything, while we know that a medical scalpel and pliers have different functions.

8.4 The uncertainty is part of the future

As mentioned at text introduction, the world has changed since introducing the computer in many applications, expanding the processing power, the processors sizes and increasing, and more and more, speed and memory capabilities.

By the cloud processing and data storage, each machine becomes a node capable to connect, access and mobilize an extreme processing power by modern applications using parallel and distributive computation.

At the present time, processors are placed everywhere in terms of IoT, and its time to improve artificial intelligence to provide better attachment between machines and humans.

Many of the technologies use brutal computer force or intense exponential processing algorithms that, nowadays, do not meet the current necessities and users' expectations.

It's a challenge to understand and deliver a good level of artificial intelligence. Some famous cases are exhibited and convince the majority of users by promoting a great experience in interaction, in a closed and controlled environment. In most of cases, it is classified as "artificial ignorance."

Most of us can easily buy an artificial intelligence machine at the common market in any e-store; it will execute very well some automation, internet searches, and will turn on/off some equipment at our house or work.

A "real" artificial intelligence machine is capable to: learn, understand, and simulate a personality addressed to it through a super user. It is still not ready and as in a new field, still has a lot to evolve.

The actual technology, although is called "an artificial intelligence machine" with woman's name; it is a door to cloud processing, representing a stage processor equipped with interfaces and sensors. This reserves a capability to upgrade internal programs and a versatility in terms of improvement in intelligence once the artificial brain that provides intelligence, is a constant machine learning process, is placed outside the equipment, at cloud computation and virtual servers services.

In terms of security and privacy, it is a disaster for people that feel spied upon by electronic devices and for whom technology can be very uncomfortable. This kind of equipment can also be a target for hackers with the ability to control all of the devices connected to it, representing a challenge ignored by majority of consumers.

Although the intelligent machines are processing information by fuzzy logic, from the 1980s until now, it is becoming more and more efficient, although it still faces one problem. The actual machines are emulating fuzzy processing and not running fuzzy processing on its internal hardware. It is still processing the binaries Zeros and Ones, while its capacity would have grown much more if its internal processing architecture were remodeled to run a more extensive vocabulary and process fuzziness [22].

From these points of view, it can be considered that the world is near a half-century late if it is recognized that fuzzy set theory could have been used to define Turing machines. It is difficult to judge the choices, because during that time even the computers were seen as very abstract and nonfriendly machines.

Bibliography

[1] B. R. Chabowski and J. A. Mena, A Review of Global Competitiveness Research: Past Advances and Future Directions, J. Int. Mark. 25(4) (2017), 1–24. http://dx.doi.org/10.1509/jim.16.0053.
[2] A. M. Turing, On Computable Numbers, with an Application to the Entscheidungsproblem, Proc. Lond. Math. Soc. s2-42 (1937), 230–265. http://dx.doi.org/10.1112/plms/s2-42.1.230.
[3] J. Łukasiewicz, O logice trójwartościowej [On three-valued logic], Ruch Filoz. 5 (1920), 170–171.
[4] B. Russell, Vagueness, Australas. J. Psychol. Philos. 1 (1923), 84–92.
[5] M. Black, Vagueness: an exercise in logical analysis, Int. J. Gen. Syst. 17 (1937), 107–128.
[6] L. A. Zadeh, Fuzzy sets. Information and Control, Vol. 8(3), Amsterdam, 1965, pp. 338–353.
[7] M. Wierman J., An introduction to mathematics of uncertainty, Creighton University, Omaha, Nebraska, 2010.
[8] F. Doria, Variations on a Complex Theme. Notes on complexity theory, comments on our work on P vs. NP (2015). http://dx.doi.org/10.13140/RG.2.1.2074.8240.
[9] T. Ross, J. Booker and W. J. Parkinson, Fuzzy Logic and Probability Applications: Bridging the Gap, Society for Industrial and Applied Mathematics, Philadelphia, PA, 2002.
[10] D. Mcneill and P. Freiberger, Fuzzy Logic: The Revolutionary Computer Technology that Is Changing Our World, 1a Edição, Touchstone Rockfeller Center, New York, 1994.
[11] M. Sugeno, Industrial Application of Fuzzy Control, North-Holland, New York, 1985.
[12] E. Mamdani, Application of fuzzy algorithms for control of simple dynamic plant, Proc. IEEE 121 (1974), 1585–1588.
[13] E. Mamdani and S. Assilian, An experiment in linguistic synthesis with a fuzzy logic controller, Int. J. Man-Mach. Stud. 7 (1975), 1–13.
[14] F. Lima, G. Morel, H. Martell Flores and N. Molines, Workshop for innovation in the metropolises: urban mobility study of a university campus, Proc. Inst. Civ. Eng., Munic. Eng. 173 (2018), 1–20. http://dx.doi.org/10.1680/jmuen.18.00027.
[15] F. A. Doria, Chaos, computer, games and time: a quarter century of joint work with Newton da Costa. 1a Edição, E-Papers, Rio de Janeiro, 2011.
[16] S.-H. Chen, Ranking Fuzzy Numbers with maximizing and minimizing set, in: Fuzzy Sets and Systems, North-Holland: Elsevier Science Publishers, 1985.
[17] T. Takagi and M. Sugeno, Fuzzy Identification of Systems and Its Application to Modeling and Control, IEEE Trans. Syst. Man Cybern. SMC-15 (1985), 116–132.
[18] Y. Takeshi, Stabilization of an inverted pendulum by a high-speed fuzzy logic controller hardware system, Fuzzy Sets Syst. 32(2) (1989), 161–180, http://dx.doi.org/10.1016/0165-0114(89)90252-2.
[19] C. A. N. Cosenza, An Industrial Location Model. Working paper. Martin Center for Architectural and Urban Studies Cambridge University. England, 1981.
[20] C. A. N. Cosenza et al., Localização Industrial: Delineamento de uma Metodologia para a Hierarquização das Potencialidades Regionais, COPPE/UFRJ, Rio de Janeiro, 1998.
[21] C. A. N. Cosenza et al., A hierarchical model for biodiesel plant location in Brazil, in: Proceedings of the Institution of Civil Engineers - Energy, Vol. 170(4), 2017, pp. 137–149.
[22] A. Syropoulos, Curso de Computação em Lógica Fuzzy (Short cource on Fuzzy Logic and Fuzzy Computation), COPPE/UFRJ, Rio de Janeiro, 2012.

Ioannis Kanellos and Elena Ciobanu

9 The semiotic topos of vagueness

Abstract: This chapter aims at demonstrating a rather undeniable fact: there is no field of human experience or intellectual activity that will remain unaffected by some vagueness for long. That vagueness appears as soon as we desire to know more. As an epistemic posture, it forms a "semiotic topos," that is, a unifying idea which is the origin of its fundamental expressions, understood as recurrent semiotic motifs. The analysis we propose is based on the assumption that human experience is thoroughly loaded with meaning through the intermediary of our interpretations. We distinguish and study four typical forms of vagueness, namely: imprecision, instability, indetermination, and indecision, and try to show that they are all epiphenomena of interpretive strategies. We conclude with a discussion addressing the relation of vagueness with meaning-making, doubt, and scientific postures and policies.

9.1 Vagueness, where are you? – A vague and skeptical exploration of vagueness

"The issue of objectivity often brings about confusion. Every experimentation is a question. And any question contains, in an often dissimulated form, an a priori judgment [...] The confusion in the issue of "objectivity" was to assume that there could be answers without questions and results independent of a question-asking being," Hannah Arendt argued in her monumental work *Between Past and Future* [1]. For our present discussion of vagueness, Arendt's statement will function as a principle of scientific ethics: the notion of vagueness cannot be independent from the human being who explores it. In other words, one cannot separate this kind of exploration from the mode of questioning, language, biography, objectives, social engagements, context and examining conditions of the one who is asking the question.

In fact, whenever we ponder on the ways in which we can establish our objectivity we always presuppose a universe that ensures the legitimacy of our method of constructing the truth – or, at least, some truth. This legitimacy supports our feeling of evidence, ensures the coherence of our constructions and ultimately offers us the valid means by which we can develop, control, and share our ideas. We are referring here to our conceptual frames, choices, thinking habits, norms, epistemological values, exigencies in terms of exactness – the more so when we are operating in the field of the sciences commonly called "exact." This more or less assumed appeal to our established modes of reading, interpreting and understanding dissimulates, most often in an unconscious way, a number of a priori judgments. The latter actually becomes our epistemic referents. Thus, we move inside a tradition of "good practices," which

https://doi.org/10.1515/9783110704303-009

provides us not only with freedoms and rights, but also, and primarily, with limits which are nothing more than the avatar of a wide and secular popular wisdom: "'This is where the world ends,' the blind said, having touched the wall."

In our view, this wall is represented by the notion of vagueness, which has become a redoubtable challenge to our sciences, especially since it was understood that complexity is to be found almost everywhere and that the truth is neither unique nor uniform. Is there any exact knowledge of vagueness? Can one construct a scientific conception of vagueness? In other words, is it possible to arrive at a completely non-vague description of vagueness? And this is where we go back to Arendt's statement. After all, the problem of who studies the notion of vagueness is a methodological, and maybe also logical, issue. Yet the way in which one conducts one's study becomes a deontological, decidedly ethical, issue. Which are one's preconceptions? Which are one's objectives?

The contributions to this volume modestly attempt to find ways of approaching and apprehending vagueness in the fields of several "exact" sciences. This may appear as an oxymoron: How is it possible to speak in an exact way about something that, by its own nature, rejects any exactness whatsoever? In reality, just like a doctor who cannot guarantee the result of his prescribed treatments (the patient's health) but who summons all his knowledge and experience for that purpose, the exact sciences are only exact in terms of their intention, engagement and method. We will enlarge on this further on. For now, we must keep in mind that the studies in this volume assume the adventure of vagueness while being unavoidably framed by perspectives, contexts and epistemologies, characteristic of their fields.

Nevertheless, before it can be conceived as the synthesis, generalization or abstraction of a variety of phenomena, states, situations, problems, etc., before it becomes a precise concept, formalized or not through a well-established theory, the notion of vagueness is a linguistic term. It is, before anything else, and primarily, a term of natural language. As such, it inexorably mobilizes the linguistic resources with which it is co-terminous and by means of which we think and act, both as humans and as scientific researchers.

In this chapter, we will attempt to show how the semiotic foundations of the notion of vagueness determine the ways in which we approach it. We hope at least to demonstrate that vagueness, viewed as a theoretical concept, has a necessarily vague destiny that is rooted in the origins of its first ecology. It is this semiotic ecology that fixes its limits and outlines its fundamental traits. More precisely, we will build our arguments according to the following guidelines:

I. Every scientific language adopts a significant part, if not the totality of, linguistic categories (particularly those related to time, space, causality, modality, etc.). Therefore, the linguistic rationality inserts itself, to various degrees and in various modes and intensities, in the scientific rationality that is being elaborated, to the effect that the latter is determined by the former at different levels of the whole construction and, eventually, of the resulting comprehension. We will thus illustrate, to a certain extent, the well-known Sapir/Whorf hypothesis, which states that reality is perceived, constructed, and lived through the categories of language.[1] Situated between two extreme positions – one which sustains that the linguistic genealogy of vagueness has nothing to do with its nature, and another which sustains that language completely determines vagueness – we will have to find a place for an acceptable, viable, and possibly useful concept of vagueness and to assume a certain position.

II. Objects, understood either in a simple naive way or in a scientific manner, can but share one of the fundamental characteristics of our human existence, condemned as it is to meaning: they too have a meaning that science is always trying to clarify, to render sharable and operational with a minimum of deformations, at the same time endeavoring to offer us means for acting accordingly. Thus, as "objects-endowed-with-meaning," they both internalize and externalize a part of the indefiniteness of meaning which can take various forms and which requires special norms, consents, and contexts of use in order to be able to satisfy a contract of exactitude. More concretely, specific types of reading and appropriate interpretations are required for objects to be correctly apprehended and profitably used by society. In any case, identifying something as "more or less vague" is only a result of interpretation, in the measure in which it is interpretation that ensures the connection and passage from reading to comprehension.

III. The consequence of such an argument is that it is not vagueness that is the exception, but its opposites: exactness, precision, clarity, definiteness, unequivocalness, etc. It is vagueness that is present everywhere and, most often, irreducible. Scientific constructions can only escape vagueness through authoritative judgments, both explicit and implicit, judgments that establish classes of equivalence corresponding to a programme of validation that is often of an applicative nature. Such judgments do not offer solutions to the problem of vagueness: They only offer the opportunity to accomplish its dissolution. They do not suppress it either: they only render it silent temporarily in order to make it operational.

Such an approach will allow us to understand that vagueness should not be seen as a scientific ogre but as a useful instrument in our endeavor to overcome the limitations of our scientific exactness. Poetry, for instance, can creatively operationalize vagueness, as many scientists have already acknowledged.

Undoubtedly, a global and direct approach of vagueness would be vain, if not downright impossible, unless one deals with things that would be rather expected and quite useless. It is obvious that the extreme variety of human practices will soon compromise any general, nontrivial definition by imposing exigencies of various natures, interests, and operationalizations, which betray the bias of specific domains. Yet, if one cannot talk "in general" about a general vagueness, one can instead talk

1 See [2]. Also: [3]. For a synthetic view, cf., for instance: [4].

“in particular” about a particular vagueness. Domain segmentation will serve in the absence of something better. This is the task of this chapter. In our semiotic endeavor, we will be guided by one of Wittgenstein’s most famous injunctions: If we must speak about vagueness, we must do it with the expected precision. Or else, we must remain silent [5].

Vagueness must therefore be approached within controllable frames of thought. This is the only means by which it may cease being uselessly vague. Intuitively, one highlights two basic categories, which are emblematic and more or less expected: the ontological aspect (vagueness in itself, the vagueness of the object of study and, more generally, of the “world”), and the gnoseological aspect (the vagueness of knowing subjects, the vagueness that elicits human representations). It is undeniable that it will be forever impossible for us to arrive at a clear-cut conclusion: We do not have the means by which to decide whether the world is intrinsically vague, and consequently contaminates our representations with its vagueness or whether it is because there is a sort of deficit in our own representations that we perceive as vague; something that is not so in reality. This dilemma already constitutes, through its undecidability, a sort of “original vagueness.”

This first opposition, which actualizes man’s opposition to the world, is projected on a multitude of distinctions linked with frames of reference and preference (a certain logic or a certain mathematical, informational, physical, chemical, medical, etc. universe), and this is rightly so. However, we will observe that all of these domain typologies approach the ontological aspect of vagueness only in an accessory manner, or at least not primarily. Vagueness is usually studied as a class of facts and by means of a more or less formal language that naturally tries to extend the expressive potentialities of natural language. These distinctions clearly aim at the same thing: to confine vagueness to the boundaries of a definition in order to turn it into a usable, that is an applicable, concept, which should at least be approachable, controllable, corresponding to facts and possessing a descriptive finesse able to satisfy certain previously formulated exigencies. A definition is more than welcome: It is necessary. Yet all definition sets limits and is conditioned by a certain position. Apriority thus constitutes the basis of the principles that regulate the human practice, which enables us to propose a definition of vagueness. For what reason do we prefer one definition to another? What class of phenomena does it address? What program of application do we envisage with this kind of definition?

Be that as it may, definition imposes itself in every case whatsoever. Even if it does not fulfill its promises, at least it has the merit of opening circles of experimentation for us. However, as H. Arendt also states in the text mentioned above, all experimentation contains the germs of a prophetization, which expresses a foresight, an expectation, a perspective, an anticipation, etc. The rest becomes formality, instrumentalization, technique, evaluation, judgment – all leading to the confirmation or invalidation of the prophetization. In science, we obviously work to understand and to find knowledge that should at least be reliable, if not altogether true. At the same time, we try to act: to foresee (what possible consequences will global warming have?), to anticipate (how to nourish 14 billion people in the near future?), to answer a concrete demand, even if it is of a general character (how to eradicate a global pandemic?). In other words, we try to create, produce, initiate, disseminate; we evaluate propositions, affirmations, assumptions, conjectures, postulates; or, when our problems become complex, we try to implement statistical models and to obtain simulations, which will help us find the most efficient and productive solution, with the least costs and risks. Yet, to accomplish this objective, we endlessly return to the departing point – the facts that we dispose of and their formulation or expression.

The notion of vagueness is similarly approached when it becomes a scientific preoccupation, simply because there is no identifiable science of the inexpressible or of the indescribable. In fact, a science is nothing but a collective effort aimed at the translation of an experience into expression. That is, at bringing experience into the field of the utterance and, further on, into a shared space which authorizes the action of collective intelligence, manifested through criticism, amendment, addition, or refutation. Expression cannot exist outside language. The lexical patrimony of our natural languages offers us a remarkable illustration of this fact. In most dictionaries (most notably English language ones), we will find that the definition of the term "fact" systematically draws on a proposition, that is, on a mode of expression and on the attribution of a truth value to that proposition.[2] This truth value is by no means absolute. The truth here is not metaphysical, untouchable, or forever stable. It only explains the framework of a human practice – a framework, which can support some credibility. Rather than telling us the truth, this framework shows us how things can be recognized as true by a community sharing the same practices, ideas, objectives, and interests. This is the central point of our reflection here: If the vision of "facts" is subtended by any logic, then it is legitimate for us to examine the role of language in the constitution of the nature of vagueness.

9.2 A vague typology of vagueness: the heritage of the semiotic locus

We must admit that the notion of vagueness accumulates various traits, aspects, and figures. It consequently forms a dense semantic field that may be the source of a great

2 The Oxford dictionary entry mentions the following usual meaning of 'fact': "a thing that is known to be true, especially when it can be proved", or "things that are true rather than things that have been invented". Examples in case are legion. The lexicographic lesson is highly pertinent nowadays, as we are witnessing the explosion of fake news: the credibility of a fact, that is, its relation to an a priori vericonditional frame, seems to turn into an instrument of power.

number of possible classifications. In reality, it proves difficult to make up an exhaustive list of the particular forms it can take in various fields. Nevertheless, if vagueness is a matter of meaning, it might be possible for us to identify some of its fundamental categories. We will discuss four such possibilities without pretending that their limits are accurate or that their content is always separable in any context: Vagueness applies to everything and, most of all, to itself! These basic categories will hopefully clarify to some extent the semiotic foundation of vagueness, explicitly or implicitly involved in the domain definition. Thus, vagueness will be studied as imprecision, instability, indeterminacy, and indecision.

It is interesting to observe that, in these terms, vagueness is lexically inscribed as opposition to things (which may be at the same time negated or devalued) generally understood as "natural," "ordinary," "native," "normal": precision, stability, determinacy, decisiveness. The prefix "in" in the four categories above confirms it. Yet, if an exact science indulges in vagueness, then would it not degrade itself by dwelling on this oxymoron? Still, as we will see, non-vagueness is only an illusion.

9.2.1 Vagueness as imprecision

Vagueness is first of all imprecision. This generally refers to the difficulty of identifying an object on which we concentrate our attention, to its perceived lack of clarity in terms of form and outline. It is therefore an imprecision originating in the knowing subject, whose visual perception, enhanced or not by specific instruments, addresses itself to various objects. Examples abound in everyday life and also in the observations made by the exact sciences, and the chapters in this book offer various illustrations of this fact. Obviously, vagueness-as-imprecision can also affect the other senses, as, for example, in the auditory perception of a signal, in the distinction among inferior and superior variations at a frequency or intensity level, in the separation of sources in sound mass effects, etc. We might also talk about olfactory, gustatory, or tactile imprecision. Imprecision can be also envisaged as an effect of sense combination, something that Aristotle (*On the Soul*, III, 2, 426b 8–22) designated by the concept of "common sense" (κοινὴ α'ἴσθησις)—a very original one for his time: the sensitive faculty, which ensures perceptual synthesis, the unity between the subject who perceives and the object that is perceived.[3] In this case, we may talk about the imprecision of "common sensibles" such as movement, stasis, number, size, etc. What appears to us as

3 The Aristotelian "common sense" is close to the notion of perception as we understand it nowadays. Being irreducible to the five senses, the "common sense" founds their synergic action when necessary. It does not have a determined application field or even an organ, it allows us to discern between distinct sense data and even assumes the consciousness of what we perceive. Vagueness thus affects our senses, their composition, distinction and consciousness.

immobile, for instance, may be "more or less immobile" (as a matter of fact, nothing is ever absolutely immobile, but only relatively immobile).

In a similar attempt at generalization, we can easily find examples of vagueness-as-imprecision related to the expression of psychological states, of feelings and emotions, and to everything that results from imagination, estimation, and memory. In fact, these are probably the most likely places where we can find this type of vagueness (manifested in the imprecision of feelings, dreams, desires, frustrations, discomfort). The exact sciences may find little interest in such facts and situations, but they cannot avoid the imprecision affecting other categories, among which causality, for example, do we always have a clear idea of the material, formal, efficient or final cause of a fact, especially when there is more than one cause? Can we ever accurately determine quantity, quality, relation, modality, or their subcategories?[4] We may even talk about imprecision in the field of action, where it would denote, among other things, lack of precision or exactness, lack of correspondence with an expected result, or very simply put, a deficit in certainty, reliability, or reassurance, etc. in the performance of a gesture, task, routine chore, protocol, etc., of a local action or of a program of actions. More generally, imprecision would here designate insufficient thoroughness in the carrying out of a job.

Imprecision permeates everything, just like meaning permeates our use of language. It appears as consubstantial with the nature of the human being seen as a knowing subject. Our short presentation of the great categories of thought and action above will have sufficed in our effort to demonstrate that, in all cases, imprecision emerges as an inadequacy observed in the actualization of a human practice, a practice which simultaneously requires a certain exigency in the treatment of data and an expected satisfaction in results. As such, it imposes a certain metrics and the frame of an evaluation process.

There is no imprecision in itself, but only imprecision related to an intention, a project, a code of communication or a list of specifications. For example, nobody would find imprecise the image in Figure 1, the reproduction of Monet's famous painting "Impression, Sunrise" (1872), which inaugurated the impressionist trend in the history of arts. In this painting, imprecision is cultivated and becomes constitutive of the object as such. On the contrary, everybody would be tempted to consider that the precision of the image in Figure 2 is directly proportional to its high definition.

4 For example, from a Kantian perspective, we can identify sub-categories such as these: vagueness affecting unity, plurality and totality; vagueness linked with reality, negation and limitation; vagueness referring to the substance/accident, cause/effect and reciprocity pairs; vagueness affecting the three pairs of modality, that is, possibility/impossibility, existence/non-existence, and necessity/contingency. More generally, we could talk of an a priori vagueness-as-imprecision which is outside experience, and of an a posteriori vagueness-as-imprecision, which is grounded in experience. Language draws on these categories and sub-categories. The Kantian framework may be understood as a philosophical endeavor to organize them rigorously, in a coherent system.

Figure 1: Monet's "Impression Sunrise." (From Wikimedia Commons, the free media repository (https://tinyurl.com/43rwz565; accessed February 2021).)

Figure 2: An ordinary photo. (This photo belongs to the authors of the present chapter.)

However, such responses are only the expression of a convention that is based on typical practices and standardized uses. If things usually appear in this way to us, this is due to the fact that information tends to satisfy us as it is. In fact, everything is always and necessarily imprecise. Researchers, whether they study the infinitely small or the infinitely big, the cell or the society, are aware that their object becomes ever more imprecise when they try to describe its structure in detail. It is as if the principle of uncertainty that was formulated almost a century ago by Heisenberg were affected by a generalization of scale, even at a transdomain level. It is known that an image that is apparently extremely clear, like the one in Figure 2, will be revealed as imprecise as soon as we change the metrics, the distance or the criteria of acceptation for the final display. If we raise the resolution to just a few pixels, the two images above will become equivalent in terms of precision. Yet, it is clear that they will no longer have the same signification. Conversely, a human being covered in clay would be as blind as a bat, whereas the bat itself does not feel that the objects it encounters in its everyday existence are at all imprecise. Then again, the objects encountered by a human being and those encountered by a bat do not have the same significance in their lives.

Precision and, consequently, vagueness, are always linked with significance. Since any life is defined by its modes of production and consumption of meaning, precision has a close relation to the life that we lead, or that we aspire to, in this huge and continuous economy of meaning. In the construction of our categories of thought and action, everything is constantly related to norms, values, rites of usage, habits, and life routines, be they real or desired. In other words, everything is constantly related to the practices that we appropriate and, consciously or not, obey. These practices are rooted in our a priori judgments.

In a 2009 interesting simulation, the French popular science magazine *Science & Vie* presented a series of 44 successive images of something we might naively refer to as "the real," in which scales were continuously decreased from one of 10/26 meters to one of 10/17 meters, that is, from the big infinity to the small infinity (https://tinyurl.com/y685ocrz; accessed February 2021). This kind of journey into the real envelops the biological and the anthropological. Thus, we realize that objects that are perfectly clear and intelligible at a certain level of analysis (essentially defined by distance) lose their precision when we pass from one power to another (inferior or superior). They not only become vague, but also disappear gradually, and even completely, to the advantage of other objects that emerge as being contained in, or containing, the former, and that, in their turn, suffer the same fate as the scale is changed. What is precise or imprecise is something that is considered within a fixed frame and within a particular horizon of expectation and that is dependent on a certain measurement and on an evaluation of the approximation. Actually, imprecision is the correlate of an evaluation judgment, which may or may not be satisfactory, depending on the context. Consequently, this judgment may or may not be acceptable, admissible, eligible, or simply valid.

To summarize, it is the context that decides the degree of imprecision and, consequently, of vagueness. There is no imprecise or vague object per se. Being vague does not refer to a characteristic in itself, but always to a couple (context and object) that is evaluated by a subject. This subject asks questions by reference to a certain metrics and to an acceptability threshold, which constitute his a priori judgments.

On the epistemic level, it is this notion of the threshold that constructs the levels of significance and, therefore, turns objects into entities endowed with meaning. Every object may lose its significance and identity, become imperceptible, or turn into something else, depending on the measure in which we move the threshold. We could not properly perceive the Eiffel Tower from a distance of 10^{-10} meters and neither could we do it from a distance of 10^{10} meters. We don't fall in love with someone that we see from a distance of 10 microns or of 10 kilometers. The sentence we are reading at this moment is not meaningful at the level of the letters that compose it and even less at the level of the Bezier curves that define the glyphs, which actualize it graphically.

The notion of the threshold may be conscious or unconscious (as in the example of the two photos above), chosen or imposed (in order to validate a quality control objective, e. g.). It may be natural (the sensitivity limits of a species) or cultural (becoming used to certain categories that inhibit others, the latter remaining active, however), general or task-specific (drawing a circle on the blackboard during a lecture or in order to model the trajectory of accelerated atoms).

Examples showing that human life is only possible within certain thresholds and that man lives in a space circumscribed by thresholds abound: the auditory discernment threshold in the human species does not always allow a subject to distinguish the overtones of a sound produced by an instrument. The relative sensitivity of the human ear makes a variation of 1 Pa to 1.5 Pa in the acoustic pressure equivalent to one of 0.1 Pa to 0.15 Pa. A discourse can be perfectly understood by a person and perceived as terribly unclear and obscure by another person, or in another dialogic context, or at a different moment, etc. After all, sciences are domains that construct the validity of certain methods. It is a known fact that the threshold effect defines the nature of a science itself and verifies its modes of objectification. Irrational numbers are evident and necessary for an ordinary mathematician, but they do not exist for someone who is calculating by means of a computer. Everything that escapes the effects of thresholds is not taken into consideration or is taken into consideration as noise. Thus, it becomes either nonexistent or vague.[5]

The concept of context is omnipresent and even inevitable, but it is also chaotic or, at least, difficult to apprehend. It refers to a number of aspects that are impossible to bring together in an overall perception. There is no science of the context, for the same reason for which Aristotle said, centuries ago, that it was impossible to constitute a science of matter (given his view on the form/matter opposition), since knowledge could only refer to forms.[6] In other words, context is not an object that could be studied like other objects, or at least not before another context is postulated in which it appears as an object.

In every context, we need to distinguish among the fundamental categories of our reason, categories that are generally sanctioned by language. We have already referred to them above in a fugitive manner: space, time, quantity, quality, relation, modality. We can also add aims, metrics, rules, and evaluation criteria to the list. Thus, the notion of context covers the whole arsenal that underlies our a priori judgments,

5 Interestingly, this question was raised from the very beginning of philosophy. It refers to the problem of sameness and otherness (when two beings are considered identical or different). See, for example, the categories of being in Plato's *The Sophist*. In his *Metaphysics* (1052b 25 sq.), Aristotle discusses the question of the threshold and of threshold effects and concludes that unity, just like being, has a different meaning depending on each predication (1053a 9 sq.); and that unity, considered as identity, signifies measure (1078b 34 sq.).

6 Aristotle, *Metaphysics*, books II and III, *passim*.

intrinsically linked to what was once called the fundamental categories of knowledge (i. e., what guarantees the possibility of knowledge).

Language is a system susceptible of giving meaning to a world, a meaning that is shared and participatory. Rather than conceiving language as a system of syntactical forms turning around in a vacuum, we may see it as a mechanism that represents our life experience, both individual and collective, as proven by the study of its six basic functions: referential, emotive, conative, poetic, metalinguistic, and phatic [6]. Thus, we may come to understand that meaning always eludes us and that it emerges only in the meeting with the other, as the expression of an intention and particularly as the outcome of interpretation. Consequently, vagueness is but the translation of a quasi-normal state of language and, more generally, of any semiotic system, that is, any system used to establish an economy of meaning. In reality, the sense of a unit (a word, a syntagm, a sentence, a paragraph, a text, a corpus of texts, etc.) is never given in an absolutely precise manner. If we accept to play the game proposed by Ludwig Josef Johann Wittgenstein, we will even find it impossible to say with any certainty what the word "game" signifies [7]. Its identity, which is a semantic one, is only precise within a context, be it textual or extra-textual, which establishes a dense network of references and exigencies. Moreover, there will always remain some imprecision that will be counteracted by a new interpretation. Given that humans live in a continuum, that is, in a continuous spatiotemporal referential, their cognition does not seem able to apprehend continuity unless they divide it into segments in order to render it operational, and thus, make it correspond to their life practices. They discretize their world by means of their languages. They make use of classes of equivalence. They establish meaning, but they maintain for themselves the possibility of modifying it at their own will. This is not very different from the situation in which a scientific perspective addresses itself to objects in order to apprehend their identity, that is, to perceive their meaning. They are precise as parts of a class of equivalence that decides their acceptability, which opens the way to experimentations, theoretizations, validations, and refutations. In the same way, we can enhance the precision of an object as much as we want, but its definitive identity and meaning, understood globally, will remain forever imprecise. Further on, we will see how this state of things is at the root of indecision.

This systematic importation of an imprecision peculiar to language into scientific objects, to the point where they become dependent on it, has preoccupied many thinkers devoted to the "divine clarity" that was much desired by some poets.[7] Would it be possible to liberate the concepts of scientific objects from this inherited imprecision? In the 17th century, Gottfried Wilhelm Leibniz tried to use his *caracteristica*

7 Arthur Rimbaud, *Une saison à l'Enfer*, "Adieu": "Mais pourquoi regretter un éternel soleil, si nous sommes engagés à la découverte de la clarté divine, loin des gens qui meurent sur les saisons ?" ("But why regret an eternal sun, if we are engaged in discovering the divine clarity, far from people that die with the seasons.")

universalis to unite all discourses by means of a *calculus ratiocinator*, a sort of algorithm, which leaves no room for vagueness and which would be able to express any discipline with the clarity and certainty of a simple calculation. In the 19th century, Friedrich Ludwig Gottlob Frege even planned to set up a language and a form of writing that would serve science and that would exclude all ambiguity. This is the famous *Begriffschrift* (*ideography*: concept writing or concept notation), the formal language of pure thought, as he called it [8], a laborious project that was however rapidly abandoned when paradoxes were discovered. David Hilbert's mathematical program inherits the same desire.

The presumptuousness of such projects, which is but a manifestation of the *hubris* (immoderation) the ancients talked about, is rooted in the underestimation or simplification of the semiotic nature of things. By confounding the predicate of existence with that of identity (both of them given by the verb "to be"), we ignore a simple fact: that essence cannot be demonstrated.[8] The feeling of the real emerges as a universe of meaning that is shared; it is not the product of a syllogism that leaves behind anything that is marked by the shades of vagueness.

Imprecision is always everywhere, in any circumstance, as long as language, that is, some semiosis, is involved. To be sure, there are aseptic spaces from where imprecision has been successfully removed, as, for example, the field of calculation, which is an activity going on outside interpretation: a simple procedural execution of formal transformations that escapes the imperatives of meaning-making. More generally, however, vagueness-as-imprecision seems to constitute the norm of a being that lives inside the space of signification. This being is condemned to move forward by successive interpretations.

9.2.2 Vagueness as instability

Something is considered unstable when it moves, when it does not remain immobile, when it lacks balance, when its form, position, behavior, etc. keep changing. Instability is also seen in something that does not resist the pressures exercised on it, something that is easily displaced, decomposed, transformed when a number of factors in its environment vary, no matter how little – especially when this variation is little, actually. In a general way, therefore, instability designates that which lacks constancy and durability. Such a form of vagueness is indicated by an impressively great number of adjectives: mobile, changing, fragile, inconstant, versatile, fugitive, moving, floating, precarious, vacillating, fluctuating, volatile, etc. even uncertain. Language thus attests to our being overwhelmed by this form of vagueness, which really appears inevitable, and which must be taken into consideration, all the more so as it turns out to be massively involved in human lives.

8 Aristotle, *Posterior Analytics*, 93b 17.

First of all, we must admit that stability is at the same time instability when it comes to the subject. Instability belongs to a universe of observables, and it must constitute facts. However, the knowing subject, while he is asking a question on the stable or unstable nature of his object of study, supposes his observations not so much in time – which would be trivial – as in a certain amount of time, depending on the type of his study. In other words, he does this within a certain duration, which is a priory human. It is obvious that an infinitely small duration (for example, of the order of 10^{-100} seconds), or an infinitely great one (of the order of 10^{100} seconds) will lead to different views on the nature of stability. If our object of study were to disappear and reappear (or if it were to modify itself and then come back to its initial state), at a speed that would be beyond our observational capacities, it would be considered stable even if it is far from that. The invention of cinema is a popular example in case which no longer surprises anyone. If we add to all these the idea of vagueness-as-imprecision as we have discussed it in the previous section, we could say that something is considered stable or unstable if we can perceive it as such, within a given temporal frame and context, by using, when necessary, specific instruments. Stability is the consequence of a deformability norm that is understood as a temporal and spatial measure. Otherwise stated, stability internalizes the a priori of a class of equivalence that observes a temporal measure. This class of equivalence is usually associated to another class of equivalence that observes a spatial measure.

To say it again, what appears as stable or unstable is not the object, but a certain couple [context, object] that is submitted to an evaluation founded on a metrics and on a list of specifications concerning the acceptance of the answer. The (nonexclusive) parameter of distance (clearly a spatial parameter) which was typical in the discussion of vagueness-as-imprecision is replaced here by duration. Depending on circumstances, we could also superpose a different form of imprecision on it. Thus we could easily draw on the arguments of the previous section to show that, after all, there is no stability or instability in itself and that everything is an outcome of the practice in which we choose to situate ourselves. This practice dissimulates its a priori and its universe of expectations in our judgments and it presides on the evaluations being made. Put differently, stability is the expression of a discretization norm that is necessary for the reaching of the expected level of efficiency. It is, to some extent, the distillate of an epistemic position.

In all likelihood, the opposition between stability and instability translates two antithetical visions of the world that are to be found, in an emblematic form, at the very beginning of philosophy, in the works of such figures as Parmenides and Heraclitus. For Parmenides, everything is immobile and stable because being is perfect and infinite: it reifies the whole space. Movement is only an illusion, that is, the product of a perception, a construction of reason. On the contrary, for Heraclitus, everything is mobile and unstable, stability being nothing but a convention.[9] It would be tempting

9 Cf., for example: [9]. Or, for a more synoptic view of the two: [10, 11].

to lose ourselves in the metaphors that recuperate such intuitions in the new ideas related to the Big Bang and its modeling in an eight-dimensional space by means of some string theory. However, metaphor is often facile and deceptive. What we should more modestly acknowledge is that here we are actually dealing with yet another convention: If there is any stability, this is because we accept to put aside certain "details," judged as "nonpertinent" and "nonsignificant." If such details can be "omitted," this is due to our a priori prejudgments, which emerge from the question we ask and from the type of answer we aspire to. A discretized universe thereby results, which allows calculations, that is, the drawing of conclusions in a space that has been carefully disinfected of any ambiguity whatsoever.

In the case of vagueness-as-instability, we also encounter a situation that is inherited from the processes that are at work in language when we give meaning to our worlds. In fact, meaning is globally unstable. It only becomes stable episodically, temporarily and only within a certain context and by observing a number of dialogic conventions. The least perturbation can alter meaning. Even a nonexplicit element that is exterior to the communicational situation can do that. For example, two readings of the same thing, at different moments, can lead to different meanings. Even when the couple [context, object] is maintained, more extended data that would contain, for example, our initial context could alter the initially constructed meaning. The analysis and interpretation of surveys is an example in case. In medicine, a diagnosis could be completely changed in the light of more global health data such as the patient's history, genetic inheritance, family pathologies, etc. Thus, there is nothing immanently stable in meaning. Everything depends on the reading that we make and on the interpretation that we apply. It is interpretation that chooses the context and, therefore, the protocol of stability. In the same way, an object appears as stable in an exact science depending on how it is looked at and on what elements are selected in order to understand it.

What we are touching on here is nothing less than the interpretative foundation of our scientific objects. If our objects are what they are, this is because we prefer, at a given moment, to follow a certain reading strategy and to interpret them consequently. The question we ask about them supposes a process of validation, either theoretical or experimental. We choose certain elements, we eliminate others, we relate everything to classes of equivalence, which decide what will be viewed as "the same" and what will be viewed as "other." Finally, we summon a number of elements, which are linked to our epistemic posture (our a priori judgments) and we constitute our object. This object will have internalized both the complex structure of these determinations and the fragility of the global construction.

To summarize: An object is stable if it can be viewed as a fixed point for a set of meaning modifications, which bestow on it a certain interpretative robustness.

In Plato's *The Sophist*, the categories of "change" and "rest," of "sameness" and "otherness" surround that of "being."[10] These are the "highest kinds." It would obviously be audacious for us at this moment to initiate an analysis of these concepts, of their multiple interpretations and posterity. Yet we feel their subtle influence in the epistemology of vagueness as we have been developing it so far (as imprecision and as instability). Instability may be essentially linked to the change/rest opposition, which is obviously connected to the duration of the observation, whereas imprecision seems closer to the sameness/otherness opposition. Simple experiences known as "time lapses,"[11] and which abound on the internet, exemplify the a priori of the temporality that is at the basis of our conception of stability.

To conclude, whenever the knowing subjects search for means to confront and usefully operationalize the instability-vagueness of their objects, it is temporality that constructs the order and the levels of being. This temporality is ultimately linked to the temporality of human life, that is, to its practices, usages, and rites, and thus, to prerogatives of efficiency.

9.2.3 Vagueness as indetermination

"All texts circulate one within another," Denis Diderot stated in *D'Alembert's Dream*, a delirious dialogue written in 1769.[12] "All is perpetual flow, every animal is more or less human, every mineral is more or less a plant, every plant is more or less an animal, there is nothing precise in nature." This is a statement that continues an already long tradition and that mirrors the same aporia in the sciences of nature: we cannot draw any clearly demarcating line between the animal and the vegetal realms or, sometimes, even between the vegetal and the mineral ones.

What we are dealing with here is less the space and perception of objects or the time and our ability to perceive them in a unity without fluctuations, but the indetermination produced by the contiguity of beings, which is probably a direct corollary of the globally continuous being. The concept of continuity has been a major conquest in the history of the development of the human spirit and of its sciences. In practice, when confronted with continuity, humans make use of their rationality, that is, they

---

10 *The Sophist*, 254d sq. Moreover, several centuries later, Plotinus, whose philosophy was based on interpretations of Plato's ideas, sustained, in his *6th Ennead* (book II) that matter, being constantly moving, does not have any permanence.

11 Cf., for example, the site www.youtube.com/c/TemponautTimelapse/featured (accessed February 2021) which proposes numerous examples of phenomena that are apparently "stable", but no longer appear so, when the temporal measure of the observation session is extended and then highly "compressed", in order to fit to our reading habits.

12 *D'Alembert's Dream* (1769; published 1830). For instance, in the Penguin Classics collection, 2000. Another good translation can be found on the internet at: https://tinyurl.com/y2jcf7q7 (accessed February 2021; translated by Ian Johnston).

build an epistemic regime founded on their languages and on their predicative capacity. They typically advance through massive, superposed, and interrelated discretizations. Aiming to render one's life meaningful, one constructs a series of orders that take the form of determined structures, even if this is done temporarily.

Determination supposes analysis and synthesis, measure, and evaluation. It is better known in sciences as "characterization" or "specification" and it refers to the fact that a word (the determined or *definiendum*) is rendered through another word or group of words (the determinant or *definiens*). Science, not to mention logic, has been preoccupied with this question for millennia. To say it again: understanding vagueness as indetermination is directly linked with language, in the measure in which the determination of an object of study generally passes through a qualification, that is, through a series of predicates that establish a sort of semantic equivalence between the two members (definiens-definiendum). In fact, determination seems to be a universal linguistic category, that is, it is rooted in a need of expression shared by all known languages.

There certainly exist other extra-linguistic forms of determination (theoretizations, modeling, experimentations, calculations, etc.), which appeal to specialized languages and, especially, to any kind of data whatsoever. Eventually, however, everything takes, or may take, a predicative form. In any case, we say things in order to give them an identity, that is, to understand them and render their signification sharable. Human rationality is always modeled *on* and *by* its language, a language that operates on three levels:

a. the level of the overall *functional system* with its general categories (time, space, genre, aspect, modality, etc.);
b. the level of *sociolects* (the languages of the social groups within which people want to live); and, finally,
c. the level of one's own *idiolect* (the language that identifies one as speaking and thinking subject, a language constituted by personal lexical, syntactical, morphosyntactical, stylistic, rhythmical, semantic, etc. choices, which transcribe one's own experience of life in society).

Thus, indetermination indicates vagueness as a failure of definition, that is, of limitation. Human beings do not live in an infinite space – as countable as this infinity may be. They need boundaries, shapes, demarcation lines, perimeters, frames. Yet they also need separations, distinctions, complete forms, barriers. In order to be able to cope with the constant entropy in which they are invited to survive, and to survive collectively at that, they have their life overflown with rules, measures, and principles. We could not live long or efficiently in the confusion of a generalized irrationality, that is, in contexts without rules, orders, principles, laws, conventions, routines, protocols, constraints, canons, or models, without a number of norm definitions, enunciated or enunciable in an explicit manner. Determination expresses a global tendency to set limits to the chaotic infinity of our worlds. We determine things in order to define them, that is, to set borders and render them manageable, to bestow an identity on things and on the state of things. We determine things to give them not only a meaning, but also a meaning that could be useful for our living-in-the-world and that could be shared with others.

Yet our determinations are rarely complete, univocal, finished. Nevertheless, this failure of definition which produces vagueness is not always the effect of a lack, it may also be the consequence of an excess of alternative determinations, which does not allow us to choose among the various possible options. From this point of view, indetermination comes closer to indecision.

It is easy to demonstrate that every object has an infinity of traits, and because of this, it could never be absolutely and totally determined. This otherwise classical problem has preoccupied philosophers ever since antiquity, who designated it by using terms like "essence and accident," "necessary characteristic," "the constitutive character of being," "the universal structure of knowledge or of consciousness," etc. At least, we usually try to find the necessary and sufficient determinations that will uniquely specify our object of interest. However, to know an object is to know exactly how it behaves or can behave in every context, which is obviously impossible, since the notion of context is itself imprecise and undetermined. In any case, it points toward an infinite multitude of conditions: states of things, situations, dispositions, modalities, circumstances, etc. To put it differently, it is not possible to determine an object completely while it is within a context that remains forever undetermined.

Here, we encounter again the basic semiotic principle that permeates every moment of our lives: meaning is never immanent to a form; context has an essential role in the determination of the meaning of any entity whatsoever. More exactly, it is the relation between context and form that makes it possible to determine an object in the best possible way or, at least, to generate an impression of determination that would be sufficient in pursuing a project of comprehension and, possibly, of action. This project, which dissimulates our a priori judgments, becomes a screen on which the shadows of a conquered, if ephemeral, determination are projected. It may suffice to fulfill the conditions of our experimentation and eventually of our evaluation, but it remains insufficient for the affirmation of the exact nature of our object of study.

We could rapidly illustrate our thesis with examples from outside language. For example, the impossible determinations in physics, like the wave/particle duality, or Heisenberg's principle of indetermination, which stipulates that we cannot always win in the field of determination by enhancing our knowledge, given that there is always a sort of natural indeterminacy, which prevents the exact simultaneous measure of two conjugated dimensions. What is more, this indetermination turns out to be structural: It does not depend on the experimental protocol and seems to constitute a determination limit of any measuring instrument. This forces us to acknowledge an original vagueness in the field of physics. Further on, quantum physics, confining itself to the salutary modesty of a mode circumscribed to what is observable, now admits that observation influences the observed system. In fact, in the process of measuring an observable, a quantum system will modify its state. This is inherent to the measure that is applied and does not depend on the experimenter's efforts not to "disturb" the observed system. Our objects will forever be undetermined, that is, they will always appear as vague.

Vagueness is a challenge for reason. The overcoming of our conveniences, generally acquired *in* and *through* our daily experience, and particularly the projection of our rationalities in the spaces of the infinitely big or small, unsettle our visions of what is precise, stable, and determined. Our grandfather Socrates already knew that he did not know anything… Or at least, that he could never know anything for sure. How can we really come to know a world that cannot be completely determined?

In a more programmatic way, this universal infusion of vagueness through indetermination in our worlds could lead us to imagine an extension of the logical approaches of the question, by projecting its modeling in intentional logic universes. We should remember here that intentional logic turns an attentive eye toward the possible worlds where our predicates could be evaluated. Yet this is another story, which is far beyond our present scope.[13]

The most intriguing aspect of vagueness-as-indetermination is probably its relation to complexity. Just like its two preceding versions, this species of vagueness subjects us to a sort of contingency thinking whose most prominent trait is the relativization of its norms. These must also be conceived as contextually sensitive, which means that they can be norms only within a certain context. Yet vagueness-as-indetermination primarily confronts us with the imperative of developing a conception of complexity, as far as vagueness-as-indetermination emblematically illustrates the limits of our thought, founded as it is on experiences of simple things. In a series of articles on complexity and the necessity of a genuine conception of complexity, Edgar Morin underlines the importance of a new rationality that could parallel the one that eventually made ancient thinkers accept irrational numbers [12]. He discusses a number of limits of our thought by showing that it appears limited because it is too much accustomed to operations that respond to more or less simple situations, for example, operations of differentiation, unification, ranking, centralization, or valorization… Such a type of thought is fatally blind to the phenomena it massively deals with. Vagueness is but an avatar of complexity: it is the place where the rational and the empirical meet, where the distance between object and subject is perturbed, where the human being necessarily becomes a parameter in the tissue of scientific meaning. This is clearly shown in every attempt we make to approach problems involving the infinitely big or small, biological or anthropological aspects.

13 Both *dialetheism* and *paraconsistecy* can be seen as alternative ways of conceiving indetermination. Paraconsistency refers to situations and conditions of non exploding consequences, whereas dialetheism proposes a new view on truth which contradicts the Law of Non-Contradiction (LNC). Cf. https://tinyurl.com/3quse6od (accessed February 2021). Nevertheless, in our analysis here, we thoroughly accept the validity of LNC.

Can this illusion of determination, which ostracises vagueness, be the domination of the simple in our scientific destinies?

It is not impossible.

In his work *Esquisse d'une Sémiophysique*,[14] the great mathematician and philosopher René Thom raises a question that is crucial and pertinent to what we are discussing here. Supposing that a naive observer contemplates a "spectacle" of forms evolving in time, Thom wonders in what conditions one could attribute a meaning to what one sees. The answer he finds is based as much on the Aristotelian *Physics* and *Organon* as on the Catastrophe theory, among whose precursors he was. This answer definitely contributes to the development of a new "metaphysics of complexity," by postulating entities of two forms: on the one hand, stable and salient ones, in opposition to their context, and on the other hand, dynamic and prehensile ones, immaterial entities, which possibly assume the function of some efficient cause. At least, they ensure the interrelation of forms and contexts of occurrence. Obviously, it is not by chance that his reflection rapidly deviates toward language, where it searches foundations for this decidedly recurrent question: how do we generate stability, even provisionally, when we are evolving in a space of dynamic entities of all types and orders which meet and determine one another incessantly?

Determination institutes and even constitutes the completeness of the objects of thought. It confers on them an integral nature, proper to our reason that is in a perpetual search of clarity and certainty. By determining something, we establish its finitude, completion, integrity, and probity. Thus, it is possible to include in this class of vagueness various phenomena of "completed incompleteness," that is, forms that we insist to perceive as complete in our effort to attribute an identity and a determined, precise, and stable meaning. Concretely, this refers to our tendency to consider certain forms complete by adding ad hoc information, which does not exist originally but whose addition is necessary if we want to find a less vague meaning, as in the well-known studies from the Gestalt theory (cf. Figure 3).

Consequently, vagueness-as-indetermination results from a voluntary act often based on experience or, at least, on former knowledge, which aims at completing a defectuous determination in order to attribute an identity to an object. In the example above, by perceiving a rectangle or a repetitive motif we have already established a meaning and, therefore, an identity, always by virtue of a series of prejudgments.

To summarize: Indetermination, understood as an emerging phenomenon of vagueness, invites us to become aware not only of our systematic interpretative operations and of our a priori, but also of the limits of our rationality. Nevertheless,

14 1988, InterÉditions. Unfortunately, the book has not yet been translated into English. Its title could be translated as "An Introduction to Semiophysics". Interestingly, R. Thom begins his first chapter (p. 15), entitled "Saillance et Prégnance" ("Salience and Prehensility"), by discussing the problem of a priori judgments.

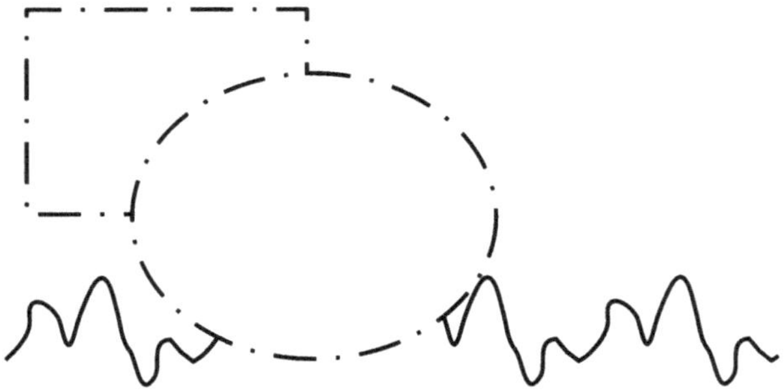

Figure 3: When the lines are completed, we perceive a rectangle that does not necessarily exist, we imagine a repetitive motif that disappears behind the ellipsis, etc.

we try to determine things by reaching a threshold, which allows our readings to acquire significance. Before this, there is vagueness; after this, we can also end up in vagueness. In fact, by trying to determine more, we summon new contexts that bring about other forms of indetermination. This form of vagueness productively translates Pascal's metaphorical formulation of the relation between knowledge and ignorance. We would naturally think that, by enhancing knowledge, every determination reduces ignorance and, consequently, renders our universes less vague. Yet this is only an illusion: the more we enhance our knowledge, the more ignorant we are. Pascal compared the quantitative growth of scientific knowledge to a sphere whose volume naturally grows as man passes through centuries, adapting himself and labouring to understand the world. What is beyond the circumference of this sphere is the space of the unknown. Nevertheless, the more the volume of knowledge grows, the more extended the surface of the contact with the unknown becomes. This also applies to vagueness-as-indetermination. Every gain in determination augments the surface of the contact with vagueness which equally grows, in an analogous manner.

9.2.4 Vagueness as indecision

Indecision is a situation that solely depends on the knowing subject's understanding and reason. It may result from a cause whose origins are to be found in one of the forms of vagueness that we have discussed so far (imprecision, instability or indetermination) or it may be the cumulative effect of their conjunction. However, it may also constitute an autonomous event, independent of these other forms of vagueness. We may actually encounter cases of indecision even when the situation seems to be perfectly clear, stable and determined. This form of vagueness designates a certain uncertainty, also a hesitation, involved in the making of a decision, maybe even a sort of voluntary suspension of judgment. Indecision brings about various forms of embarrassment, given that it plunges the subject into confusion, doubt, and perplexity, it makes him vacillate and imposes on his vision of things various opposite movements without resolution, incessant fluctuations, all comparable to what we have already seen in the previous cases of vagueness.

The uncertainty with which it envelops both the comprehension and action of the subject goes hand in hand with a new parameter, which engages the perception of values on the part of the subject: risk awareness. This risk is inherent to the consequences of decision-making. Thus, this type of vagueness is instituted through the intervention of the subject whose indecision grows directly proportionally with the perceived importance of the risk involved. Besides, the notion of risk is a complex one and it involves numerous fields, forms, and degrees. It refers both to the specification of the danger and to the evaluation of its gravity, both to the probability of occurrence of an undesirable situation and to the cost of its acceptability. Differently put, indecision systematically summons elements and factors that are outside the very situation that generates it, in the measure in which they are all related to the subject and his/her context. Indecision is not a concept that could be constructed out of some elementary bricks, by following a compositional mode (bottom-up). It is instead an epistemic posture tightly connected to exterior data. The "decision risk" is necessarily semiotised, that is, it is necessarily embedded in a space of semantic values that are in effect in a certain society – be it only a circle of scientists preoccupied by this kind of risk. Obviously, all these exterior factors blur the perception of the studied object which thus escapes the usual knowledge legitimation procedures. Finally, this object appears as vague inasmuch as it has inherited the indecisiveness of the subject.

A few examples.

In the following situation (see Fig. 4), which consists in a generalization of an example that we find in R. Thom's book,[15] we are required to decide which of the (b), (c), and (d) curves represents a phenomenon whose evolution could ideally be represented by (a) and which remains, of course, unknown.

The three models are clear: They are not affected by imprecision, instability, or indetermination. It is, however, impossible to decide because the correct answer depends on the a priori judgment associated to the question – a judgment which privileges certain horizons of expectation. If we were to decide which curve best and genuinely represents the phenomenon from a qualitative point of view, this would be curve (b): It has exactly the same appearance and form, with the exception that it is positioned at the bottom of the figure. From a quantitative point of view, this would be curve (c), even if it has nothing in common with the form of our object of study, which it patently betrays; nevertheless, it has the merit that it is closest to curve (a) if we take into account the standard deviations from the ideal representation of our phenomenon. Finally, on the "economic" level, it is curve (d) that is preferable as it is easier and less costly to produce, and it mediates between quality and quantity (it

15 *Op. cit.*, p. 21.

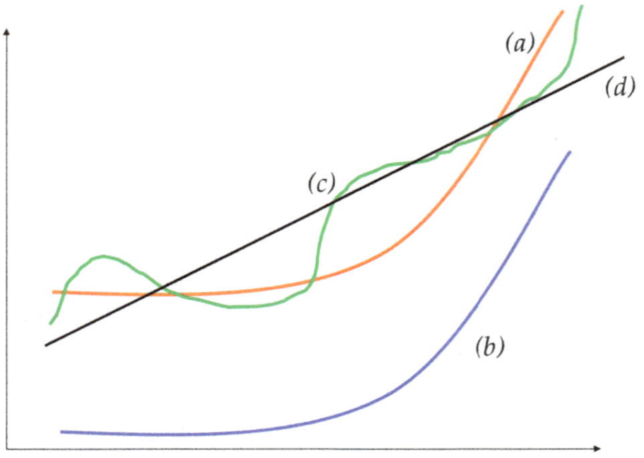

Figure 4: Looking for the best representation.

respects the ascending appearance of (a) and it offers a convenient approximation in terms of positive and negative differences). The situation is not vague in itself: It becomes so due to nonexplicit judgments dissimulated in the question that is asked. In other words, in such cases, vagueness is a derived product, which results from an insufficiently clarified consensus at the beginning.

The situation described above is not much different from that of daily personal or institutional decisions, or even that which concerns the "nonexact" sciences. For example, how could we answer this question: Which is the country that was the most rewarded at the Olympic Games in 2008? Figure 5 illustrates a certain nonvague representation (issued by the Olympic Games committee) of the first twenty countries, which made history at this event organized in Pekin, China [13].

If we read this table in a noncritical manner and, furthermore, if we accept its data as "objective," we will only reveal our lack of awareness of the a priori that are contained in the initially asked question. Indeed, if we only count the golden medals, the winner will be China. If we take into consideration all of the medals, or only the golden and the silver medals, the winner will be the USA. If we take into account the weight of obtained medals (as a coefficient of importance), the results will totally depend on this. Thus, possibilities are unlimited: For instance, supposing that the golden, silver, and bronze medals have weights of 3, 2, and 1, respectively, the USA continues to be the winner, whereas if we consider that the weights are 10, 5, and 1, China will be by far the undeniable winner. By linking these same results to the population, GDP, HDI, gender of the athletes, life expectancy, population aging, budget allotted to sports, etc., things become ever more uncertain. Yet we make our decisions in an entirely different way. For example, Jamaica will be the winner if we prioritize the criterium related to the degree of sports involvement among the population, with its 1.67 medals per million of inhabitants (as compared to 0.03 in China, and 0.11 in the USA). We could interpret

Rank	NOC	Gold	Silver	Bronze	Total
1	China (CHN)*‡	48	22	30	100
2	United States (USA)‡	36	39	37	112
3	Russia (RUS)‡	24	13	23	60
4	Great Britain (GBR)‡	19	13	19	51
5	Germany (GER)‡	16	11	14	41
6	Australia (AUS)	14	15	17	46
7	South Korea (KOR)‡	13	11	8	32
8	Japan (JPN)‡	9	8	8	25
9	Italy (ITA)‡	8	9	10	27
10	France (FRA)‡	7	16	20	43
11	Netherlands (NED)	7	5	4	16
12	Ukraine (UKR)‡	7	4	11	22
13	Kenya (KEN)‡	6	4	6	16
14	Spain (ESP)‡	5	11	3	19
15	Jamaica (JAM)‡	5	4	2	11
16	Poland (POL)‡	4	5	2	11
17	Ethiopia (ETH)‡	4	2	1	7
18	Romania (ROU)‡	4	1	4	9
19	Cuba (CUB)‡	3	10	17	30
20	Canada (CAN)‡	3	9	8	20

Figure 5: 2008 Summer Olympics medal table, by the Olympic Games committee.

the previous conclusion by saying that in Jamaica the Olympic ideal is 53 times more significant than in China (or 15 times more significant than in the USA), etc. Things become even more "vague" in terms of decidability when we combine these criteria. The huge literature of the discipline called "decision-making" attests to this fact. It is not that decision is impossible, but that it is fundamentally uncertain, as long as its a priori are not explained. This case is very similar to that of ambiguous figures and, more generally, to that of perceptual illusions, and it is based on a choice of relevant elements in our reading of the data.

The example above offers us a clearer view of the relation between vagueness and the interpretative processes. However, one may not like sport or the Olympic Games – after all, they are not claimed as objects of study by any exact science. In that case, we might alternatively think about physics and the forms of indecision due to the participation of the observer to the phenomenon and its evaluation, as in the famous thought experiment proposed by Erwin Rudolf Josef Alexander Schrödinger in which one cannot decide whether a cat is or not alive.[16] Yet again, we might think about the wave/particle duality and its relation to the observer's context. The most typical case of indecision is obviously encountered in complex contexts in which competing dynamics force us to display our models of simple causality for the sake of an aggregate conception of small causalities, that is, in which a great number of infinitesimal causes lend undecidable, because chaotic, aspects to the respective effects.

In the exact sciences, just as in other sciences, indecision is but a residual incapacity that translates the difficulty of deciding, which interpretation is the correct one. Let us allow ourselves a harmless digression into the realm of poetry.

Indeed, we may experience a similar thing when we try to interpret a number of lines in a poem. For example:

> *Dying*
> *Is an art, like everything else.*
> *I do it exceptionally well.*
> *I do it so it feels like hell.*
> *I do it so it feels real.*
> *I guess you could say I've a call.* [15]

Poetry is often detested precisely because of its semantic indecision, which compromises any easily identifiable certainty, clarity, invariability, concreteness, or stability. However, the indecision encountered in poetry is not of the same nature as that which ultimately renders scientific interpretations vague. Vagueness, as we have been discussing it so far, is based on a movement that is opposite to that of what literary criticism commonly refers to as "poetic ambiguity." In his widely read *Seven Types of Ambiguity*, William Empson [16] astutely analyzed various manifestations of this phenomenon in the work of a number of canonical English writers and proposed a sort of classification of the forms that poetic ambiguity may take by complexity and depth. "Ambiguity," the English critic argued, "can mean an indecision as to what you mean, an intention to mean several things, a probability that one or other or both of two things has been meant, and the fact that a statement has several meanings."[17] If readers assume the difficult task of deciphering poetic ambiguity, whether it involves words, sentences, or discourses, they will be richly rewarded, because, instead of getting farther from meaning, he will experience semantic revelations, which may significantly compensate for the perpetual loss that comes through the inherent and inescapable vagueness of language.

The lines quoted above from Sylvia Plath's famous poem *Lady Lazarus* create ambiguity by intentionally unsettling the reader's expectations at several levels. The po-

---

16 Cf. [14]. In this book, the author begins by exploring, in his first chapter, the concepts of possibility, probability and context. In section 1.3.3., he launches the idea that probability is situated between ontology and epistemology. We can easily extrapolate this notion to vagueness.

17 *Op. cit.*, pp. 5–6.

etic persona defyingly declares her "exceptional" ability to perform death as an "art" in lines that oppose a nursery rhyme musicality with a dark irony that theatrically engages the other/reader in an uncertain dialogue/game. In a culture which relegates suicidal acts to the realm of the undesirable or blameable, this type of discourse dares to affirm the truth of an extraordinarily gifted, but intriguingly tormented mind. The strange beauty of the lines and the elegance of the utterance, associated with an unconventional subject matter enrich the possibilities of interpretation, and thus reveal how aesthetics is intertwined with the epistemological and ethical dimensions.

Poetry tells us something about the world and about ourselves in ways that differ essentially from those of science, but it is not the less valuable for that. Quite the opposite, actually: "when it comes to atoms," the physicist Niels Henrik David Bohr intriguingly observed, "language can be used only as in poetry. The poet, too, is not nearly so concerned with describing facts as with creating images and establishing mental connections." This brilliant scientist argues that "we are suspended in language," and considers "the division of the world into an objective and a subjective side much too arbitrary; the fact that religions through the ages have spoken in images, parables, and paradoxes means simply that there are no other ways of grasping the reality to which they refer. But that does not mean that it is not a genuine reality. And splitting this reality into an objective and a subjective side won't get us very far" [17].

Semantic polysemy may appear as problematic and deceptive in the scientific process of interpretation. Nevertheless, it becomes the most significant instrument by which we can transcend the limits of our language and knowledge. In poetry, this transcendence happens not only by recreating meaning in order to get closer to the inexpressible ontology of things and beings, but also through the subtle challenge addressed to those very aprioricaI judgments with which we approach it in an attempt at comprehension. If this is so, it is also because, beyond its disturbing semantic innovations, poetic discourse is unavoidably affected by the same vagueness we have been focusing upon in this chapter.

The density of poetic meaning and the effort involved in the apprehension of its semantic "plus-value" usually intimidates readers who prefer "simple," concrete, possibly "exact" narratives. After all, exact sciences are nothing but objectification-driven prose. Yet, "any prose statement could be called ambiguous."[18] The propensity to certainty that discourse seems to manifest is only an illusion: The decidability of meaning is only an exception. It is always the result of a great effort of clarification and explanation of the context and of the intention of the reading strategy and, therefore, of the values which institute the object as such and qualify the shareability of a piece of knowledge.

18 W. Empson, *op. cit.*, p. 1.

Table 1 resumes such type of "ambiguous prose" on a cognitive science experiment. It presents the answers of 7 observers confronted with a simple task: the detection of the presence of a particular signal (S = signal) systematically produced in a noisy environment (N = noise). The subjects were asked to say whether the signal was actually present (N and S) or not (only N). The first column registers the times when the observers perceived the signal while it was actually present, while the second column includes the times when the observers believed to have perceived the signal while it was actually absent.

Table 1: Who is the best observer?

	100 attempts — Condition: N+S	100 attempts — Condition: only N
Observer 1	45	3
Observer 2	50	11
Observer 3	84	3
Observer 4	76	3
Observer 5	30	3
Observer 6	6	0
Observer 7	42	18

It is obviously easy to make comparisons among the observers who had the same performance in the second column (the third observer is better than the fourth, who, in turn, is better than the fifth). However, the situation is generally undecidable: it is impossible to say who the best observer is. Specifically, it is impossible to decide between observer 3 and observer 6. Just as in the discussion on the winning country at the 2008 Olympic Games, everything depends on the risk that we associate with a bad answer, that is, with the cost of an erroneous answer. This may lead us to decide that the best observer is 6, despite the fact that he did not detect the signal in 94 % of the cases in which it was actually present [18].

Once again, observation, all by itself, plunges us into uncertainty as long as we haven't explained the context in which it is made, a context which triggers the risk that retrospectively blurs the phenomenon itself. We believe that we describe some objectivity. But, indeed, "Everything we call real is made of things that cannot be regarded as real" and the issue "is not about how the world is but what we can say about the world."[19]

Wherever we turn, it is language that sets the limits of our rationality and even the horizon of the utterable.[20] Situations are similar, whether we are talking about a cat whose fate depends on the observer's action, about the detection of a signal, or

19 N. Bohr, *op. cit.*

20 In the stoic philosophy one may find a similar conception in the notion of "λεκτόν" ("utterable").

about the declaration of a victory in sport or in an election campaign. It is not much different when we have to decide why (or how)"dying is an art," which articulates hell and reality. After all, a phenomenon is like the Lord of Delphi: he neither speaks nor conceals, but signifies. To make a decision about what it offers us as information, that is, about its meaning, is to make a decision about our questioning and reading modes. Finally, it is all about the construction of an answer to a question that addresses the foundations of a number of a priori, which it assimilates and sets in motion.

9.3 Discussion and conclusion

In this chapter, we have identified four concepts by means of which we have tried to approach, develop, and even structure the notion of vagueness in a series of category schemes. What we have proposed is only one alternative among others. We have tried to redefine the concept of vagueness through some of what we considered to be its principal forms. Simultaneously, we have endeavored to fight against our own spontaneous realism which ceaselessly pushes us to interrogate the space that we need to give to our objects in a way that "naturally" leads us from the ontological to the epistemological level. In fact, nothing is more obscure or more intriguing than being, just as nothing is more precarious or more doubtful than our knowledge.

It is always possible – and even legitimate – for us to begin by reflecting on the traditional, intuitive, and pedagogically effective categories that are grounded in the clear opposition between subject and object. After all, their equally evident reconnection happens through knowledge. We will never be able to know with any certainty whether knowledge originates in the object or in the subject or whether it is independent, autonomous, and somehow ideal and primordial, as the Platonic tradition generally asserted. We have to admit that the notions of subject and object are comfortable and impose themselves as empirically clear notions. Therefore, in the absence of any means by which to decide if stones, living beings, galaxies, and angels swim in the same vagueness as we do, it seemed reasonable –or at least justified – for us to approach vagueness as a peculiarity of human consciousness. This does not reduce, however, the richness of possible alternatives for studying vagueness: It only gives us an opportunity to develop it in relation to the human.

It remains completely possible to imagine an approach that would concentrate on other topics, not necessarily coextensive with those that we have proposed, which would undoubtedly require adapted rationalities and dedicated forms of logic at the same time. For example, we could distinguish a vagueness of perception, of observation, of instrumentalization, of modeling, etc. We could equally identify a vagueness of reading or of interpretation, which would be definitively subordinated to a vagueness of comprehension. Furthermore, we could conceive a recurrent or systematic vagueness, which would resist scientific methods and experiments, or a circumstantial and transitional vagueness, which would be linked to the state of our sciences in a certain age. We might even come to talk about vagueness as the correlate of an episteme, as Paul-Michel Foucault (best known as Michel Foucault) defined it. We remember that for this author the episteme designates the totality of the modes of production (and consumption) of knowledge in a given age. We might add an episteme of calculation which nowadays supports various integrations and uses of the numerical technologies in our practices and brings about a remarkable change in our relation to knowledge, by stimulating a radically different manner of conceiving the relation between knowledge and power [19, 20]. Thus, the list of possible approaches of the concept of vagueness is probably open forever.

Our objective has been to demonstrate a simple and undeniable fact: No field of human experience or intellectual activity remains unaffected by vagueness. While we restored the relation between vagueness and the subject by bringing vagueness closer to human experience, we found it only natural to envisage it as "being-in-time" rather than as an absolute "being," that is, as something that is grounded in the lived human experience as constituted within societies, rather than as an abstract, more or less transcendental, essence. This position allowed us to avoid the disastrous investigation of vagueness conceived as a separate autonomous entity. In order to pursue this necessarily limited project, we have also summoned a priori judgments to answer our questions. The following sections synthesize and discuss some of their implications. Obviously, we are aware that, in writing these last lines, we will not be able to avoid vagueness ourselves.

9.3.1 Vagueness and meaning-making

Studied in connection to experience, vagueness accompanies every human presence and its manifestations: from the length of one's hair to the sequencing of some genetic code, from one's estimation of the distance to a galaxy to its representation in a poetic line, from one's perception of one's desires to the principles of a political system. The extraordinary lexical richness of languages when it comes to the expression of vagueness already underlines its profusion in our lives. Seen as imprecision, instability, indetermination or indecision, vagueness grows, multiplies, and propagates itself. It soon becomes approximation, indistinction or inexactness. Depending on contexts, objectives and means, vagueness generates in the knowing subject weakness, hesitation, confusion, uncertainty, imperfection, trouble, ambiguity, instability, versatility, fragility, precariousness, inconsistency, suspension of judgment, indeterminacy, conditionality, embarrassment, indecisiveness, perplexity — or their opposites. In any case, it denotes an uncomfortable comprehension. It claims an important part of doubt.

Everything — absolutely everything — is vague, even the sentence that one is reading now. However, vagueness cannot exist on its own, autonomously. It colonizes our existence as rational human beings. Vagueness is a sort of virus of our knowledge universe, that is, an exiled entity which needs a host in order to live. This host is human experience, on which vagueness inscribes its code, thus producing a multitude of approximations in our manner of understanding. Our immune response is only provisional: It takes the form of a reaction on the part of knowledge, which quickly exudes something precise, stable, determined, certain, or operational in a given situation. Prisoners of our world and enthralled by it, we may not always clearly see that everything is relation and that what we understand as precise, stable, determined, or certain only describes our relation to a collective experience, in a certain society and history. That is to say, we are conditioned by our intention to produce and consume meaning, to think and act together *with*, *for* and *like* the others. It is this experience that imposes its criteria of acceptance on our answer. Yet this answer does not solve the problem of vagueness: only that of the conditions of its acceptance. The main medium for the modeling and transmission of the cultures and knowledge issued from this experience is represented by our language. This explains our choice of a semiotic approach of vagueness.

Being inherent to language, vagueness cannot be transcendental, as it cannot exist outside experience. It can neither provide us with any conditions of possibility for our knowledge – quite the contrary: It appears as a universal constant of human experience; more precisely, of our daily experience of meaning. In other words, vagueness is not something that descends from the skies, but something that precedes the semantic economy inside, which we conduct our destinies. It thus feels natural to conclude that everything is contaminated by vagueness, given that human meaning is always and everywhere vague. Nevertheless, this does not trouble our daily lives in any excessive measure: we have tamed the vagueness of phenomena in the same way in which we tame the vagueness of the meaning that we exchange at every moment and in every circumstance.

Such an idea has obvious affinities with the phenomenological tradition and its fundamentals as we find them described by the historical representatives of this trend of thought [Edmund Gustav Albrecht Husserl, Martin Heidegger, Maurice Jean Jacques Merleau-Ponty (best known as Maurice Merleau-Ponty), etc.], and also, a long time before them, by the neoplatonic or even the epicurean philosophers: the questioning of objective research, the importance of what appears in the experience of a social and historical subject, the replacement of the essence of things by the meaning of things as parts of the human universe, the interrogation of evidence related to intersubjectivity... In our present approach of the concept of vagueness, these modes of being have been intentionally subordinated to the primacy of human semiotic action. Our analysis is based on the assumption that human experience quickly and radically appears as linguistically semiotised, that is, loaded with meaning through the intermediary of our language. Thus, by means of our interpretations, the structure of our language will sooner or later constrain our readings and determine our understanding of the objects, states, situations, or worlds that we set out to examine.

By what right can we say something particular about vagueness? Obviously, by the right of language, that is, by the right to make meaning through the plumbing of the resources offered by the language that we speak and that encodes a collective experience which precedes and accompanies us. This language, through its general system, sociolects and idiolects, models our reason by aligning our cognition and our understanding with those of the others with whom we socialize. In reality, it is through language that vagueness becomes a social universal. It may further become a political fact.

In what space do we talk about vagueness in one way or another? Decisively, we do it in the space of language, which offers us dogmas, transcodes our rationalities, frames, orients our perceptions, and accompanies us in our conceptual adventures. "Dogma, rigidly imposed by tradition, crystallizes in formalism," Edward Sapir argued. "Linguistic categories constitute a system of dogmatic wrecks, and they are dogmas of the unconscious."[21] The Sapir/Whorf hypothesis is never very far from the worlds that we endeavor to construct.

A semantic approach of vagueness would undoubtedly remain insufficient. Even an exhaustive list of notions synonymous or related to the concept of vagueness could not completely cover its ecologies and possibilities, as they will soon display a systematic semantics of a classical nature. Differently put, the concept of vagueness is not limited to its relations of opposition, contradiction, implication, or neutrality resulting from an analysis derived from the famous logic square developed by Algirdas Julien Greimas.[22] The very omnipresence of vagueness renders vague even these relations and the square itself! Our analysis is semiotic —not semantic: It is motivated by the need of a genuine hermeneutics of vagueness. This would actually be an unusual hermeneutics, because vagueness, devoid as it is of a territory of its own, and permeating everything, also applies to itself: There is no precise, stable, determined, or decidable vagueness. Vagueness is definitively vague.

This search for a "lost hermeneutics" of vagueness would be founded on an observation that we have already formulated above: Establishing meaning always sets in motion interpretative processes. Language offers us the frame for our categories of understanding, but it is interpretation that finally decides whether something (an object, a state, a situation, or a phenomenon) is vague or not. It is interpretation that can make something appear as nonvague and give it the appropriate form for integration in a scientific discourse. It is interpretation that can operate the choices which truncate vagueness in order to give it an exploitable form that would be definitely acceptable because sufficiently precise, stable, determined, and decidable. Language is

---

21 See [21]. In this book we find the premises of what subsequently became better known as the Sapir/Whorf hypothesis, which is, however, only a reformulation of the same conception manifested by W. von Humboldt and many other thinkers in the wake of Hegel. For a more recent study, see, for example, [22].

22 For example, [23]. Or, in a more synthetic and critical manner: [24].

our sine qua nonframe. Interpretation is the main vector in our semantically saturated lives. Therefore, if vagueness is the fundamental trait of our being-in-the-world, its negotiation and operationalization are simultaneously achieved through interpretation.

9.3.2 Vagueness and doubt

Vagueness, as we have seen, is related to doubt. Yet this should not be understood as a simple disguise of Cartesian doubt. The concept of vagueness expresses a methodological doubt. It appears as soon as we formulate our insatisfaction with the validity of our certainties, as soon as we think it necessary to reconsider the a priori that subtend our certainties, as soon as we rethink the mode of acquisition of what we judge as very precise, clearly stable, completely defined or absolutely certain. The consciousness of vagueness does not prevent the progress of knowledge: It only gives it form, let us say a different "mode of being," by relating it to its horizon of appearance. It is obviously a consciousness that invites the return to things and, at the same time, to oneself. This is not less valuable knowledge, but only knowledge that serves other projects.

This sort of doubt that lurks in vagueness reminds us that we cannot have a total experience of the world, that there will always be something that is beyond us, imperceptible for us, at least temporarily, or that there will always be fields to explore beyond what is obvious. To be a subject is not only to have a world, but also to have a horizon associated with this world. We cannot perceive this horizon in its totality, as it continually moves with us. It is not a place to reach, but a mode of arrangement of our spaces of existence and, in particular, of our objects, which will never be more than what we ask them to be, that is, objects of response and satisfaction.

Thus, in a certain way, vagueness could be understood as the expression of the transcendental conditions of doubt – what allows doubt to exist and operate not for blocking knowledge, but for refreshing it by renewing our scientific paradigms. What is important is not the "what" of this vagueness-as-doubt, but its "how": this "how" invites us to abandon, if momentarily, our conceptions of identity so that we could open ourselves to an epistemology of similarity.

9.3.3 Vagueness and sciences

Vagueness is a question-word, not an answer-word. It is an enigma. It is not an object of study like others: it articulates a refutation by which we conceive our science, given that it encapsulates the project of objectification that we assume. Yet, this project of objectification is only the expression of our decidedly interpretative relation to our worlds. It might be useful to think about the concept of interpretation that produces or destroys a case of vagueness as a criterion for the classification of sciences and also for the qualification of their protocols of objectification.

If things appear as precise, stable, definite, or decidable, this is because we interpret them in a certain way, and not in another, that is, because we have chosen to approach certain aspects and not others, to establish our criteria and thresholds of acceptation, to eliminate various "details" that trouble us, to restructure, to add elements, to reconstruct the whole. It is by means of such a process, which gives meaning to our objects of study, that we have also analyzed vagueness in this chapter. Interpretation is a strategy aimed at presenting things with the precision that we want. "The precision that we want" here refers to "the exigency of a society preoccupied by the same problems and which may consider them as receivable." Therefore, sciences would no longer appear as exact or nonexact, hard or soft, etc.: They would only be sciences that give different functions, values, and permissions to interpretation and, consequently, to the vagueness of their objects.

The elimination of vagueness cannot be done by means of a technique, method, or theory, or even by means of the radical change of a scientific paradigm, but by decision. Vagueness cannot be eliminated, only disguised, by dispossessing it of its causes. By doing this, we place our objects within manageable frames. Clarity is but the price of a reduction, which is, at the same time, a reduction of the meaning of our objects. The exact sciences actually oscillate between "fuzzification" and "defuzzification." Fuzzification designates processes which deviate what appears firm, clear, and sharp into something fuzzy or fuzzier. Defuzziness designates the opposite trajectory going from something fuzzy to something crisp and disambiguated. Both introduce new linguistic variables or limit the possible ones. Defuzzification is only a methodological principle necessary for a transitional epistemic period, during which efficiency is generally obtained by freezing the meaning of the objects under study for a limited period of time, in a given context and for an established objective. This is part of the standard objectification work in the exact sciences.

In fact, objectification often (or maybe always) contains a part of the consensus which eliminates what produces vagueness. The obtained objectivity is only the situation in which a meaning has been constructed that must imperatively appear as founded. It is a meaning susceptible of being included in discourse and of becoming sharable, a meaning that should be in keeping with the reality of a community, possibly serving some purpose. Most of all, it is a meaning coherent with the system of objectivities erected before or around it.

Once more, we see how this mode of action opens the door for the subject. This subject, however, is no longer the individual subject, but a social one – the exponent of a collective intelligence. Thus, vagueness reveals how the subject "penetrates" the object. It also reveals the way in which the desire to know imprints itself on the studied object, the whole operation being sanctioned by the authority of a community, which grants the rights of simplification and validates the results obtained through that simplification. From this perspective, we are tempted to say that vagueness translates the entrance of a socialized epistemology into the field of the ontological.

Beyond the ontological dimension, which seems destined to remain forever impenetrable and undecidable for us, we must identify an epistemic vagueness whose objectivity is no longer a matter of individual taste or appreciation, and which escapes simplicity ("I believe that this is or is not vague"). It is, instead, an intersubjective construction ("for these reasons, whose validity we acknowledge, and of which we are aware when we are asking our questions, we believe that this is or is not vague"). This feeling of clarity that permeates our exact sciences, this conviction that we have vanquished vagueness in a number of fields, is only the fragrance of a discreet consensus, the child of a shared authority. It is not true, or even partially true, but significant and useful for scientific thought, completely analogous to a law or norm that regulates the modus vivendi of a community that asks the same questions as us.

In the light of the concept of vagueness, the difference between the exact and nonexact sciences might be seen like this: The former identify vagueness but immediately search to eliminate it through a consensus that promises efficiency, utility, and correspondence to theories and experiences; the latter simply recognize and accept vagueness, and evolve within its boundaries. In the exact sciences, the regime is somewhat totalitarian: vagueness is subjected. It is not tolerated. It is excluded or rejected so that one could proceed further and come to fulfill the list of requirements in terms of performance and productivity. In the nonexact sciences, one is subjected to vagueness, but this comes at the price of a complexification of the action which often becomes difficult to define and even more difficult to quantify.

Be that as it may, vagueness is intertwined with complexity in both cases. Thus, in both of them, vagueness invites us to interrogate the classical triangular structure of "subject, object, and knowledge," which is becoming ever more obsolete nowadays; and to prefer another in which consensual subjects replace the subject, and in which the object identifies itself with its contextual meaning, and in which knowledge is the result of an intersubjective investment of the interpretation in the objects under scrutiny.

The real and, maybe, only problem of vagueness is that it cannot be what it is, but something else that is linked with scientific policy and probably with the ambition and role of sciences and technologies in the transformation of our world: Finally, who is the master of vagueness? Who decides what will be considered vague or not and how? Vagueness will always remain an epistemological issue, but its nature will henceforth concern the normative structure of a community which shares the same interrogations and exigencies and which transforms it into an intersubjective – therefore interconceivable – concept. It will no longer be a matter of individual decision or the universal imposition of an apriorical exteriority, but the result of an intersubjective consensus.

R. Thom used to define an object as the invariant of a series of transformations. His idea is very interesting. The feeling involved in the observation of an object that does not appear at all vague presupposes a number of transformations, which do not alter its identity. Thus, by changing this series of transformations, it is possible for us to render our objects less precise, less stable, less definite, or less certain. The decision to change the transformations announces another project of research or another question, or, again, another manner of making science. We have tried to show that our objects become differently sensitive to contexts as soon as their identities are conceived in terms of their meanings.

Vagueness appears as soon as we desire to know more. It is similar to a noisy zone toward which our interest is suddenly directed. In that zone, we hope to find a new meaning. Vagueness constantly sends us back to the consciousness of our methodological imperatives. As such, it cannot constitute a doctrine. "What is well known is only generally so, it is precisely because it is well known that it is unknown," Georg Wilhelm Friedrich Hegel argued [25]. Some of us might overhear, behind the German philosopher's voice, the whispering of a smiling Socrates for whom the sensible world will forever lack intelligibility. Vagueness may be the deep cause of this state of things. Among other things, it may reveal our anxieties while we endeavor to reintroduce some lucidity in the universe of our sciences.

Bibliography

[1] H. Arendt, Between Past and Future—Eight Exercises in Political Thought, The Viking Press, New York, 2006, pp. 1961–1968, Penguin Publishing Group, ISBN: 978-1-101-66265-6.
[2] E. Sapir, in: Selected Writings of Edward Sapir in Language, Culture, and Personality, D. G. Mandelbaum, ed, University of California Press, 1983.
[3] B. Whorf, in: Language, Thought, and Reality: Selected Writings of Benjamin Lee Whorf, J. B. Carroll, ed, MIT Press, 1956.
[4] J. A. Lucy, Linguistic Relativity, Annu. Rev. Anthropol., 26 (1997), 291–312. https://www.jstor.org/stable/2952524?seq=1 (accessed February 2021).
[5] L. Wittgenstein, Tractatus logico-philosophicus, Proposition 7: "Whereof one cannot speak, thereof one must be silent" ("Wovon man nicht sprechen kann, darüber muss man schweigen"), 1921.
[6] R. Jackobson, Linguistics and Poetics, 1960, https://tinyurl.com/y3pobmg3 (accessed February 2021).
[7] L. Wittgenstein, Philosophical Investigations, § 66.
[8] https://gallica.bnf.fr/ark:/12148/bpt6k65658c (accessed February 2021).
[9] N.-L. Cordero, Les deux chemins de Parménide, Vrin, 1997 (ISBN: 978-2-87060-011-5).
[10] J. Barnes, The Presocratic Philosophers [Revised Edition], Routledge Taylor & Francis Group, London & New York, 1982. (ISBN: 978-0-415-05079-1).
[11] D. W. Graham, Heraclitus and Parmenides, in: Presocratic Philosophy: Essays in Honour of Alexander Mourelatos, V. Caston and D. W. Graham, eds, Ashgate, Aldershot, 2002, pp. 27–44. ISBN: 978-0-7546-0502-7.
[12] E. Morin, On complexity, Hampton Press, 2008, ISBN: 978-1572738010.
[13] https://en.wikipedia.org/wiki/2008_Summer_Olympics_medal_table (accessed February 2021).
[14] M. Bitbol, Mécanique Quantique: une introduction philosophique (Quantum Mechanics: a philosophical introduction), Champs Flammarion, 1996.

[15] S. Plath, Lady Lazarus, in Collected Poems, Faber & Faber, London, 1981, p. 245.
[16] W. Empson, Seven Types of Ambiguity, Chatto and Windus, London, 1949.
[17] The Philosophical Writings of Niels Bohr, Vol. 2 (1932–1957): Atomic Physics and Human Knowledge. https://tinyurl.com/4x8mxvwt (accessed February 2021).
[18] Y. Baumstimler, Peut-on parler de décision dans la perception?, Bull. Psychol. 23(1–3), N° 280, (1969), 56–62.
[19] M. Foucault, The Order of Things: An Archaeology of the Human Sciences (1966), Routeledge, 2001.
[20] M. Foucault, The Archaeology of Knowledge. And the Discourse on Language (1969), Routeledge, 2002.
[21] E. Sapir, Language. An introduction to the Study of Speech, 1921.
[22] L. M. Ahearn, Living Language: An Introduction to Linguistic Anthropology, John Wiley & Sons, 2011, ISBN: 978-1-4443-4054-9.
[23] A. J. Greimas, Structural Semantics: An Attempt at a Method, Nebraska Univ. Press, 1983.
[24] L. Hébert, The semiotic square, https://tinyurl.com/y5xpgkuo (accessed February 2021).
[25] G. H. F. Hegel, Phenomenology of Spirit (transl. by W. V. Miller). Oxford Univ. Press, 1976.

Index of notation

Index of people

www.ingramcontent.com/pod-product-compliance
Lightning Source LLC
Chambersburg PA
CBHW082034230326
41598CB00081B/6344
* 9 7 8 3 1 1 0 7 0 4 1 8 1 *